COOPERATION IN THE
ENERGY FUTURES OF CHINA
AND THE UNITED STATES

AF148439

National Research Council

Chinese Academy of Sciences

Chinese Academy of Engineering

NATIONAL ACADEMY PRESS
Washington, D.C.

NATIONAL ACADEMY PRESS • 2101 Constitution Ave., N.W. • Washington, D.C. 20418

NOTICE: The project that is the subject of this report was approved by the Governing Board of the National Research Council, whose members are drawn from the councils of the National Academy of Sciences, the National Academy of Engineering, and the Institute of Medicine. The members of the committee responsible for the report were chosen for their special competences and with regard for appropriate balance.

The project was also approved by the Chinese Academy of Sciences and the Chinese Academy of Engineering.

This study was supported by Grant No. DE-FG01-98EEE35047 between the National Academy of Sciences and the U.S. Department of Energy Grant Number C826453-01-0 from the U.S. Environmental Protection Agency and funds provided by the U.S. National Academies. The Chinese Academy of Sciences and the Chinese Academy of Engineering each provided funds to cover costs of their participation in this study. Any opinions, findings, conclusions, or recommendations expressed in this publication are those of the author(s) and do not necessarily reflect the views of the organizations or agencies that provided support for the project.

Additional copies of this report are available from National Academy Press, 2101 Constitution Avenue, N.W., Lockbox 285, Washington, D.C. 20055; (800) 624-6242 or (202) 334-3313 (in the Washington metropolitan area); Internet, http://www.nap.edu

International Standard Book Number 0-309-06887-8

THE NATIONAL ACADEMIES

National Academy of Sciences
National Academy of Engineering
Institute of Medicine
National Research Council

The **National Academy of Sciences** is a private, nonprofit, self-perpetuating society of distinguished scholars engaged in scientific and engineering research, dedicated to the furtherance of science and technology and to their use for the general welfare. Upon the authority of the charter granted to it by the Congress in 1863, the Academy has a mandate that requires it to advise the federal government on scientific and technical matters. Dr. Bruce M. Alberts is president of the National Academy of Sciences.

The **National Academy of Engineering** was established in 1964, under the charter of the National Academy of Sciences, as a parallel organization of outstanding engineers. It is autonomous in its administration and in the selection of its members, sharing with the National Academy of Sciences the responsibility for advising the federal government. The National Academy of Engineering also sponsors engineering programs aimed at meeting national needs, encourages education and research, and recognizes the superior achievements of engineers. Dr. William A. Wulf is president of the National Academy of Engineering.

The **Institute of Medicine** was established in 1970 by the National Academy of Sciences to secure the services of eminent members of appropriate professions in the examination of policy matters pertaining to the health of the public. The Institute acts under the responsibility given to the National Academy of Sciences by its congressional charter to be an adviser to the federal government and, upon its own initiative, to identify issues of medical care, research, and education. Dr. Kenneth I. Shine is president of the Institute of Medicine.

The **National Research Council** was organized by the National Academy of Sciences in 1916 to associate the broad community of science and technology with the Academy's purposes of furthering knowledge and advising the federal government. Functioning in accordance with general policies determined by the Academy, the Council has become the principal operating agency of both the National Academy of Sciences and the National Academy of Engineering in providing services to the government, the public, and the scientific and engineering communities. The Council is administered jointly by both Academies and the Institute of Medicine. Dr. Bruce M. Alberts and Dr. William A. Wulf are chairman and vice chairman, respectively, of the National Research Council.

CHINESE CHAIR

Prof. Lu Yongxiang * CAS, CAE — President, Chinese Academy of Sciences

CHINESE MEMBERS

Prof. Cai Ruixian, CAS — Institute of Engineering Thermophysics, Chinese Academy of Sciences

Prof. Fan Weitang, CAE — President, China Coal Society

Prof. Hu Jianyi, CAE — Research Institute of Petroleum Exploration and Development, China National Petroleum Corporation

Prof. Wang Yingshi — Institute of Engineering Thermophysics, Chinese Academy of Sciences

Prof. Yan Luguang, CAS — Institute of Electrical Engineering, Chinese Academy of Sciences

Prof. Yao Fusheng, CAE — Bureau of Machinery Industry

Prof. Zhao Renkai, CAS, CAE — China National Nuclear Corporation

Prof. Zheng Jianchao, CAE — Electric Power Research Institute of China.

Prof. Zhou Fengqi — Director, Energy Research Institute, State Development Planning Commission

CHINESE EX-OFFICIO MEMBERS

An Jianji — Associate Director-General, Bureau of International Cooperation, Chinese Academy of Sciences

Cao Jinghua — Director, Office of American and Oceanian Affairs, Bureau of International Cooperation, Chinese Academy of Sciences

Liu Xiaobei — Secretary for Foreign Affairs, Chinese Academy of Engineering

CHINESE STAFF

Cui Zhengxin — University of Science and Technology of China

---

*Chinese names are listed family name first, given name second.

v

Acknowledgments

We wish to thank the U.S. Department of Energy, the Environmental Protection Agency, the U.S. National Academies, the Chinese Academy of Sciences, and the Chinese Academy of Engineering for their financial support of this project. The committee also wishes to thank Lawrence Berkeley National Laboratory for hosting a meeting of this group, and the Sierra Pacific Power Company for a visit to the Piñon Pine Power Project. Finally the committee wishes to thank all the individuals consulted over the course of preparing this paper, especially those of the Energy Information Administration; these informal consultations were an invaluable component of this committee's work.

We would like to recognize the contributions made by Inta Brikovskis, Staff Officer at the NRC, and those of Feng Liu, an independent consultant to the NRC. Douglas Bauer, Executive Director of the Commission on Engineering and Technical Systems at the NRC also deserves recognition for his work in helping conceptualize this project and in forming the committee.

This report has been reviewed by individuals chosen for their diverse perspectives and technical expertise, in accordance with the procedures approved by the NRC's Report Review Committee. The purpose of this independent review is to provide candid and critical comments that will assist the NRC in making the published report as sound as possible and to ensure that the report meets institutional standards for objectivity, evidence, and responsiveness to the study charge. The content of the review comments and draft manuscript remain confidential to protect the integrity of the deliberative process. We wish to thank the individuals listed below, who are neither officials nor employees of the NRC, for their participation in the review of this report. While the individuals listed below have provided many constructive comments and suggestions, it must be emphasized that the responsibility for the final content of this report rests entirely with the authoring committee, the NRC, and the Chinese Academies of Sciences and Engineering.

U.S. Report Review Committee

Linden Blue — Vice Chairman, General Atomics
William Chandler — Director, Advanced International Studies Unit, Battelle Memorial Institute
Richard Cooper — Boas Professor of International Economics, Harvard University
Elisabeth M. Drake, NAE — Associate Director for New Energy Technology, MIT Energy Laboratory
Robert Frosch, NAE — Senior Research Fellow, Harvard University
Donald L. Guertin — Director, Program on Energy and the Environment, The Atlantic Council of the United States
Kent F. Hansen, NAE — Professor of Nuclear Engineering, MIT
Edwin E. Kintner, NAE — Retired Executive Vice President, GPU Nuclear Corporation
John W. Landis, NAE — Retired Senior Vice President, Stone & Webster Engineering Corporation
Michael May — Research Professor, Engineering-Economic Systems & Operations Research, Stanford University
David H. Pai, NAE — President, Foster Wheeler Development Corporation
Richard S. Stein, NAS, NAE — Goessmann Professor of Chemistry, Emeritus, University of Massachusetts

This report was also simultaneously reviewed by a Chinese Report Review Committee. We wish to thank the Chinese reviewers listed below who provided many significant comments and suggestions for the report.

Chinese Report Review Committee

Prof. Pan Jiazheng, CAS, CAE — Vice President, Chinese Academy of Engineering
Prof. Wu Chengkang, CAS — Institute of Mechanics, Chinese Academy of Sciences
Prof.Yang Bailing — Vice President, Chinese Academy of Sciences
Prof. Zhang Kan — Director General, Bureau of International Cooperation, Chinese Academy of Sciences
Prof. Zhu Xuan — Secretary General, Chinese Academy of Sciences

Preface

Adequate supplies of energy at reasonable costs are essential for long-term economic prosperity and social development, and to correct past environmental problems as well as to prevent new ones.

The energy futures of China and the United States are intimately linked: both countries draw on the same international sources for imported oil and are affected by changes in its price and availability, and both countries will be depending on similar energy technologies and will jointly benefit from technological advancements. Energy production, conversion and use has significant environmental consequences, and these consequences in both countries have similar potential to benefit and to harm.

Environmental consequences from energy production, conversion, and use also provide a linkage between the two countries: similar technologies and practices can be used to prevent pollution; damage in each of our countries is of interest to the other because we share a common world heritage and concern for our peoples. Moreover, environmental consequences can be transboundary—for example pollution of the shared oceans and atmosphere; and they can also be global, as in the emission of greenhouse gases.

The two countries share some of the same challenges, for example: improving energy efficiency; increasing the contribution of cleaner energy supply to lessen local, regional and global pollution; reducing the harmful effects from coal combustion; increasing the efficiency of petroleum extraction and use; and increasing the use of renewable energy resources. However, they have complementary and mutually supporting abilities to contribute to meeting these challenges. Each country has experience in specific technological fields. Each country has advantages in implementing the next stage in technological development—for instance, China may provide the opportunity for installation of next-generation nuclear power facilities or clean coal technologies, and its unserved rural regions may be ideal for proving next-generation renewable distributed power systems such as solar. In comparison, the United States has a longer history in electric generation and can develop the techniques and practices for improving reliability and efficiency of power plants and for extending their useful life. Consequently, the two countries can make more progress working together, using each other's strengths, than they can working separately.

China and the United States will play an important role in the world's energy future. Together they comprise just over a quarter of the world's population and account for over a third of total world energy use. Both have maintained strong and growing economies in spite of a global downturn and the Asian economic crisis. The United States ranks first in energy consumption and production and is responsible for about 24 percent of the world's energy-related carbon emissions. China is second in energy consumption and accounts for 13 percent of global energy-related carbon emissions. In addition, China is a leader among the industrializing countries of the world and the United States among the industrialized. Their example will be studied and may be emulated around the world.

Our conclusion is that cooperation on energy matters is in the best interest of both the United States and of China. Further, this cooperation will have beneficial effects on other countries as well. Therefore it is vitally important for the two countries to develop and sustain the kind of intellectual, economic and political relationships that will foster cooperation in our long-term energy futures.

In a January 1997 meeting of the Presidents of the NAS and the CAS, Dr. Alberts and Dr. Zhou proposed that a cooperative program on energy be undertaken by the four Academies (NAS, NAE, CAS, and CAE). Members of four Academies and several energy experts were organized to implement this report which is aimed at identifying initiatives and providing recommendations to both governments. It is our hope that the result will be helpful to both China and the United States, as well as to other countries.

The group from the four Academies responsible for this study of challenges and cooperative opportunities in the energy sectors of China and the United States has examined the likely trajectories of each country's energy development from the present to the year 2020. Given the long life cycles of energy infrastructure, the decisions made by our respective governments—and the cooperation between them in the period discussed in this study—will have implications reaching far beyond the next few decades. Cooperation in both the short and near term can make a critical contribution to a sustained energy future for our nations who, being the largest countries and economies in the developed and the developing world, set an example for other nations.

The paper which follows is a consensus document. All members of the U.S. and Chinese committee have agreed to each of the conclusions and recommendations.

Contents

Executive Summary

CHARGE TO THE COMMITTEE

As we enter the next millennium, with the world's population and economic expectations still growing, we recognize that today's energy resource utilization likely will undergo profound changes early in the next century. The importance of such a transition should not be understated. Our nations must cooperate to provide clean, affordable energy for economic growth and social development, and work to minimize future energy security concerns, environmental threats to our global society, and the health and economic impacts of energy production and use. This report focuses on collaborative opportunities between China and the United States—two large and influential nations—but the lessons learned here also can serve the larger global community.

Both countries share the common goal of prosperity, peace, and good environmental quality for citizens of all countries and future generations. Energy and its utilization are core components of pursuing this goal and are the focus of this joint committee work. Current and projected trends in resource production, distribution, and use pose uncertainties and challenges for the near, medium, and long terms. Collaboration with regard to technology, policies, and institutions can help to make the uncertainties less threatening, and help to overcome the challenges. The objective of the current effort was to identify both challenges and opportunities where collaboration on the accelerated development and deployment of energy technologies and appropriate institutional innovations can contribute to meeting the goals of both countries. In particular, the committee has worked to identify mutual benefits that can be derived from jointly pursuing some of these opportunities.

The Academies of Sciences and Engineering in China and the United States convened this joint *Committee on Cooperation in the Energy Futures of China and the United States (CCEF)* to address these issues.

NATIONAL CONTEXT FOR THIS REPORT

The primary audiences for the committee's work are the governments of both countries, whose collaboration in science and technology, when initiated in 1978, became the foundation for meaningful interaction between the United States and China. The importance of this interaction has not been forgotten, nor has it diminished in the past two decades; on the contrary, more collaboration in science and technology between the two countries is being undertaken now than ever before.

The reason for this increase in international cooperative efforts has become apparent in recent years: each government understands that the actions taken within its national boundaries have impacts far from its borders. This is apparent with the globalization of our national economies, the global nature of technological innovation, the amount of attention being given to environmental change and resource sustainability, and the worldwide nature of the environment.

In the course of their interaction, the two governments have faced many difficulties, a reasonable expectation given the breadth of interests and views. The relationship between China and the United States is currently in a period of some tension over both political and economic issues. However, it is the firm belief of this committee that our common interests and the solutions to our problems lie in better understanding and an increased interaction between the two nations.

This committee worked to better address common needs and concerns in the energy sector. The committee conducted its work shortly after a series of meetings at the highest political levels in both governments, and this report attempts to capture the spirit of those interactions and assist in institutionalizing the many opportunities present in the energy sector.

CHALLENGES AND OPPORTUNITIES

The United States and China share a primary challenge of providing—in a period of projected growth—adequate and reliable energy services in both the near and long term while minimizing adverse health, economic, and environmental impacts associated with energy production and use. Components of this challenge include:

- growing oil dependence, and in particular, increasing dependence on petroleum imports;
- adverse local, regional, and global economic, health, and environmental implications of energy-related emissions, particularly from coal use;

- proliferation, cost, safety, and waste management concerns associated with use of nuclear resources; and
- economic barriers for developing and deploying new clean, efficient technologies and renewable energy sources.

China faces additional challenges, both as a developing country and by virtue of its geography and natural resources, including:

- limited developed energy resources per capita and an inaccessibility of energy resources—this is particularly true in the abundance of coal and hydropower resources far from population centers;
- limited use of high-quality energy resources, particularly the lack of access by about 40 million people to commercial sources of electricity;
- an incomplete energy infrastructure, especially for electric power and natural gas; and
- high energy intensity—the energy use per unit of economic output; and
- limited availability of capital to meet growing energy needs.

Challenges specific to the United States include:

- the uncertain future of commercial nuclear power;
- the absence of political consensus on the need for, and means to, address global climate change, especially in light of the potential retirement of U.S. nuclear power plants; and
- growth in energy use and emissions in the transportation sector, due to larger vehicle fleet and increased vehicle use.

Cooperation—governmental, academic, scientific, nongovernmental, commercial, and financial—offers opportunities to address these challenges through:

- increased use of energy-efficient technologies currently available;
- moving to cleaner, more efficient combustion of fossil fuels;
- increased use of higher-quality fuels, including electric power;
- expanded research, development, and demonstration (RD&D) of technologies;
- mutually beneficial and collaborative transfers of capital, knowledge, and technology; and
- collaborative economic and environmental initiatives that benefit mutual national interests and those of the global community.

STRUCTURE OF THE REPORT

The report is divided into three chapters: Chapter 1 is a sector-by-sector overview of the current and projected energy developments for both countries. It contains subsections on baseline projections for each country, variations from these baselines—through deployment of advanced technologies, creation of markets, different economic growth cases, and so on—as well as a brief sketch of existing international collaborative efforts within the energy sector. Chapter 1 is based on work done by both governments, as well as by outside groups. It is not editorial in nature, but sets the stage for the two remaining chapters.

Chapter 2 offers perspectives and commentary by the committee. It was surprising to many of the members that the challenges faced by both countries are so similar, despite the many differences in the energy sectors of each country. Also, the opportunities also have important similarities. Because the charge was to develop collaborative opportunities, the committee attempted to identify the options in the development of the energy sectors of both countries whereby each would benefit.

In the first part of Chapter 3, the committee focuses on the mechanisms by which the Academies of Sciences and Engineering in the United States and China might structure further work to provide continuing input in the crucial energy area. The cross-cutting recommendations address issues having impacts across the entire energy sector: the use of market mechanisms to accelerate the development and deployment of advanced technologies; how best to utilize existing financial institutions to increase application of advanced technologies; and research, development, and demonstration of advanced technologies. In each of these areas there are specific opportunities, for example, development of advanced renewable energy technologies, and these have been highlighted as subpoints to the broad cross-cutting recommendations.

In the sections that follow the initial cross-cutting recommendations, the committee identifies collaborative opportunities through initiatives particular to a specific portion of the energy sector. These initiatives highlight specific technology opportunities or institutional mechanisms that could be strengthened through increased collaboration between public- and private-sector institutions in both countries.

Recommendations to governments share a common principle: increased public-private collaboration. In both the United States and China, many of the technologies whose applications are desirable reside in the private sector; one of the challenges in our public sectors is to create the market conditions necessary for effective development and deployment of these technologies.

NATURE OF THE RECOMMENDATIONS

The recommendations contained in this report are made by a committee selected by the four Academies of Sciences and Engineering and are intended for institutions in both countries. In some cases, specific changes in policy are noted for a particular government, but these suggestions are being made by a joint committee.

HIGHLIGHTS OF FINDINGS AND RECOMMENDATIONS

The committee focused on initiatives to meet the challenges in the energy sector through a period ending in 2020, recommending:

- promoting investment in new technology development in both countries and the possibilities for collaborative research, development, and demonstration;
- cooperation and collaborative knowledge transfer to facilitate sound planning for cleaner and more efficient energy services in China and the United States; and
- increasing cooperation between China and the United States in the accelerated use of advanced energy technologies.

These recommendations are both institutional and technical: they address what type of market conditions are necessary to encourage deployment of advanced technologies, and the portfolio of technologies that show promise for the time period covered by this study. The rapid growth and modernization of the Chinese energy and transportation systems offers opportunities for other nations to share new technologies and to learn from the insights that come from early deployments. This can truly be a mutually beneficial collaboration.

The committee also notes some of the ongoing collaboration in the energy sectors between the United States and China and offers recommendations on high-priority areas with existing institutional relationships. As the scope of work was so broad, the recommendations developed are numerous, in some cases detailed, and cover many different aspects of the energy sector. The committee places importance on all recommendations. Some, however, deserve special emphasis and priority because of their significance to both countries. By virtue of their breadth and potential impact across the entire energy sector, all three of the cross-cutting recommendations merit this special emphasis. The second clean coal recommendation (pertaining to technology adaptations), the second energy efficiency recommendation (strengthening and expanding research, development, and demonstration of energy efficiency technologies), and the natural gas recommendation also deserve high priority.

In order to promote the cooperation between China and the United States, a primary task is to find ways to institutionalize the opportunities and initiatives. The committee believes that a long-term cooperative program in the energy area among the Chinese Academy of Sciences, the Chinese Academy of Engineering, and the U.S. National Academies could help to sustain joint programs on new and ambitious opportunities and to institutionalize the exchange of information and interaction between governments and industry. Regular communications are vital, and they represent an important component of our institutions' ability to provide scientific and technical policy advice to our governments.

CROSS-CUTTING AND ACADEMY INITIATIVES

The CCEF recommends that a standing committee be established among the four Academies to identify opportunities for research, development, demonstration, and deployment of cleaner and more efficient energy technologies. (A1) The committee identified two possible subcommittees to begin the task of structuring such an interaction. The committee also suggested that this new standing group establish an "Academies-Industry Forum" to convene top-level energy industry representatives, researchers, and government agencies from the United States and China.

Considering the national importance placed on the reduction of greenhouse gases by many countries, including the United States, and on economic development and local and regional environmental control by China, our governments should initiate a dialogue on incentive programs to accelerate the deployment of advanced energy technologies which would become cost-effective in the expected economic environment. (A2) Initial support for new and advanced technologies is necessary due to the difficulty in achieving immediate competitive economic results compared to costs of existing deployments. The committee further suggested that our governments consider collaborating on a technology demonstration project to illustrate the mutual benefits of the Clean Development Mechanism as established under the 1997 Kyoto Protocol to the 1992 United Nations Framework Convention on Climate Change.

The CCEF recommends a broad participation by agencies from both countries in energy cooperation, with financing agencies and facilities specifically emphasizing their support for energy efficiency, renewable energy, and other advanced clean-energy technologies. (A3) *Initial support for new and advanced technologies is necessary due to the difficulty in achieving immediate competitive economic results compared to costs of existing deployments.*

In a series of subrecommendations the committee identified U.S. government agencies and programs that, if authorized to undertake activities in China, could have significant benefits in both countries as well as globally. First among these is the U.S. Agency for International Development (USAID) whose activities in institutional and market reform, technical training, building and transferring experience with new technologies and management techniques could do much to further our common development goals. The committee also noted the high potential value of the activities of the U.S. Trade and Development Agency (TDA) and Overseas Private Investment Corporation (OPIC) if they were authorized to undertake activities in China.

The potential impact of these institutions undertaking collaborative work with China spans all of subsequent areas of discussion. In deliberations leading to the drafting of this report, the committee noted the political sensitivity of these recommendations but felt that this consideration was more than counterbalanced by the scale of common interests involved and by the critical impacts of the energy sector on health, the economy, and the environment.

ENERGY USE AND END-USE EFFICIENCY

The recent passage of the Energy Conservation Law in China presents a wide variety of opportunities for collaborative work. The initiatives here are intended to assist in the implementation of this important legislation.

The CCEF notes the inadequate support to date for investment and trade in advanced energy efficiency technologies between the two countries and recommends that new resources be devoted to expanding these activities. (B1)

The committee makes a subrecommendation to create or strengthen a mechanism to increase information exchange on energy efficiency—with a particular emphasis on financing mechanisms for efficiency projects. It also recommends a high-level bilateral study of institutional innovations to promote trade and investment in energy efficiency technologies.

The CCEF recommends significantly strengthening and expanding the existing program of collaborative precompetitive research, development, and demonstration of energy efficiency technologies between the two countries. (B2) The committee also endorses the activities of the Sino-U.S. Working Group on Energy Efficiency and its subgroups, and recommended expanding and strengthening this Working Group as a means of carrying out the initiatives on technology research, development, demonstration, and policy assessment.

CLEAN COAL

The CCEF recommends that the U.S. Department of Energy and the Chinese Ministry of Science and Technology review and strengthen the programs and processes for utilizing and disseminating clean coal technologies (CCTs) in China. (C1) In a subrecommendation, the committee advocates support of an independent information center and clearinghouse on CCTs and also provides preliminary topics of interest.

The CCEF recommends that both governments convene a group representing public and private interests to assess the variety of clean coal technologies, to determine their suitability for China's market, and to identify adaptations that will be required for each technology to make it more suitable for near-term use. (C2) The product of such an interaction would be a detailed research and implementation plan both to accelerate near-term use and to encourage continued technology improvement for long-term sustainability. The committee also suggests several high-priority areas with which such a group might begin their discussion.

NATURAL GAS

The CCEF recommends that both governments work collaboratively to explore possibilities in developing an overall strategy for accelerated natural gas development in China that includes production of domestic natural gas and coalbed methane, and imports of piped natural gas and liquefied natural gas. (D1) Such a strategy would address both technology and policy components, including market development, environmental reform, and financing. In all of these activities there may be a role for U.S. government entities not currently authorized to work in China (see Recommendation A3).

The committee also notes that increased activity in the coalbed methane (CBM) industry in China could speed the development of China's natural gas industry. China's experience with coalbed methane (CBM) could be valuable in the future, as the utilization of natural gas and CBM involves many of the same techniques and infrastructure, both physical and institutional.

The CCEF also recommends that China consider distributed electric power generation options using remote sources of natural gas or CBM from smaller fields to meet the energy needs of remote populations currently without access to commercial energy, and to augment existing services through increased reliability and lower total cost.

PETROLEUM

The CCEF endorses the objectives of the ongoing "Oil and Gas Forum" initiated in 1997, and recommends that the following major areas for cooperation in the petroleum sector between China and the United States be on the agenda of this continuing bilateral dialogue: restructuring issues; long-term sector strategies; exploration and resources assessment; refining technologies; transportation fuels; and environmental protection. (E1)

The CCEF found common energy security concerns that stem from each country becoming more dependent on petroleum imports (E2) and proposes that China and the United States collaborate in a comprehensive analysis of the potential and merits of national and regional strategic petroleum reserve systems.

The CCEF recommends that the U.S. and Chinese governments and industry establish a dialogue on light transport vehicles, including alternatives to petroleum transport fuels, and cooperate on both technology development and market creation. (E3)

RENEWABLE ENERGY

For commercial and near-commercial technologies the committee found that the focus of renewable energy cooperation (primarily a private sector responsibility) should be on lowering costs; for demonstration technologies the focus should be on introducing their advantages and reducing costs through scale-up and experience. The committee also sees value in cooperative efforts in precompetitive cooperative R&D for technologies expected to be of considerable importance beyond the time frame of this study.

The CCEF finds that U.S.-Chinese cooperation in the following renewable energy areas would be especially helpful: establishing a policy framework; developing technology and market assessments; strengthening research and development cooperation; and training. (F1) The committee also identifies USAID, TDA, and OPIC as having experience in renewable energy technologies and recommends that their programs be undertaken in China to further efforts in the above areas.

The CCEF recommends that both governments establish periodic reviews of renewable energy collaboration to better meet strategic objectives of both countries. (F2) A regular review of the progress of renewable energy collaborations—and frequent information exchange among all parties—would result in a better coordination of programs.

NUCLEAR ENERGY

For a sustainable energy future, preservation of nuclear energy as a power generation option is important: this is particularly so given heavy coal dependence in both countries. *The CCEF found that the following priorities for our governments concerning a commercial nuclear power program are similar: the ability to prevent proliferation of fissile materials and handle spent fuel and waste; safety in the design and operation of nuclear plants; and a desire to improve the economics of nuclear plants. (G1)* The committee recommends collaboration on the standardization of plant design and operation, and on a specific effort to demonstrate long term storage and disposal for spent fuel and high-level waste.

Our governments and industry should play a leadership role in international organizations, such as the International Atomic Energy Agency and the World Association of Nuclear Operators, to assure that international commitments, regulations, and appropriate measures are defined and implemented. (G2)

The committee emphasizes the importance of bilateral cooperation and endorses the framework of the Agreement on Intent of Cooperation Concerning Peaceful Uses of Nuclear Technology (PUNT) signed by both governments in 1997. (G3) The committee also encourages expanding programs such as the Nuclear Energy Research Initiative to include China and recognizes the need for high-quality personnel at each country's nuclear facilities.

ELECTRICITY TRANSMISSION AND DISTRIBUTION SYSTEMS

The trend in both China and the United States is toward deregulation of electric power markets. Though the two countries are at different stages, both intend to separate power generation from transmission and distribution. This intention presents an important opportunity for collaboration between the two countries. There is also the need for a closer coordination with the international financial institutions who are providing support to China in efforts to create a competitive generation market, one that includes independent power producers.

The CCEF recommends that the governments of the United States and China collaborate on measures to foster the development of a successful electric power sector, including:

*   • planning for interconnection and further development of the electric power grid*
*   • encouraging international financial institution financing for the electric power grid in China,*
*   • examining the conditions necessary to promote increased interest in independent power production in China, and*
*   • improving the adequacy, quality and reliability of electric power in China.*
(H1)

The committee also notes the importance of ongoing efforts to restructure China's energy sector and made several cross-cutting recommendations to that effect, above.

Technology collaboration—both public and private—between China and the United States could contribute to optimizing the electric power grid and increasing reliability under circumstances in which both countries would benefit. The key component in ensuring successful private sector collaboration is the formation of a transparent and competent regulatory process in which all parties have confidence. A structured exchange between the Electric Power Research Institutes in each country would provide significant opportunities for information exchange and could provide insight into advanced technology deployment in the United States especially in flexible alternating current transmission, the performance of clean coal technologies, and distributed generation deployment. Such a relationship could also provide the connection necessary to build on the experience gained in the United States in providing power to outlying rural areas.

China and the United States share an interest in developing more economically viable distributed power sources for remote areas and should identify cooperative activities that advance this interest. (H2) China's ongoing efforts to provide energy services to its large rural population provide a significant opportunity to examine the role of non-grid connected systems, especially those that incorporate a renewable energy component.

Highlights of Conclusions and Recommendations

Recommendations in *italics* are noted by the committee as having the highest level of priority.

CROSS-CUTTING INITIATIVES

- *Create an Academies committee to identify opportunities for research, development, demonstration, and deployment of clean and efficient energy technologies*

  Consider sponsoring an "Academies-Industry Forum" to convene top-level energy industry representatives, researchers, and government agencies

- *Initiate a governmental dialogue on incentives to accelerate deployment of advanced energy technologies, perhaps demonstrating a technology project under the Clean Development Mechanism*

- *Increase interaction among U.S. and Chinese government agencies and recommend that government financing agencies emphasize energy efficiency, renewable energy, and other advanced clean energy technologies*

  Authorize the U.S. Agency for International Development, including the U.S.-Asia Environmental Partnership, to work in China

  Authorize the U.S. Trade and Development Agency and the Overseas Private Investment Corporation to work in China

ENERGY USE AND END-USE EFFICIENCY

- Increase investment and trade in advanced energy efficiency technologies
  Expand information exchange, support training, and facilitate investment and joint ventures
  Undertake a high-level bilateral study of institutional innovations to promote financing

- *Strengthen and expand existing collaborative pre-competitive research, development, and demonstration of energy efficiency technologies*
  Expand and strengthen the Sino-U.S. Working Group on Energy Efficiency and add a team on transportation energy efficiency

CLEAN COAL

- Review and strengthen existing governmental programs to utilize clean coal technologies in China
  Support a private center to undertake studies and analysis, conduct seminars, and perform institution building
  Hold focused workshops and training on management practices, progress at various clean coal technology demonstrations, and how to work with bilateral and multilateral agencies

- *Convene an industry public and private-sector group to assess clean coal technologies, determine suitability for China, and identify specific adaptations required beginning with specific high-priority areas*

NATURAL GAS

- *Work collaboratively with industry to develop an overall strategy for accelerated natural gas development in China*
  Specific priority areas include exploration and resource assessments, gas market infrastructure, and environmental policy reform
  Coalbed methane priorities include extraction, market definition, and access
  China should consider distributed electric power generation options using remote sources of natural gas or coalbed methane

PETROLEUM

- The Oil and Gas Forum should focus on: restructuring and building market institutions; long-term strategies on development, operations, imports, and strategic reserves; exploration and resources assessment; refining technologies and alternate transport fuels; and environmental protection in a system that reflects real costs

• Our countries have common energy security concerns and could benefit from collaboration on an analysis of strategic petroleum reserves and macroeconomic impacts of fluctuations in the world petroleum market

• Governments and industry should establish a dialogue on light transport vehicles and alternatives to petroleum transport fuels and cooperate on both technology development and market creation
    Consider a broader examination of urban transportation systems in both countries

RENEWABLE ENERGY

• U.S.-Chinese cooperation would be especially important in: setting up a market-oriented policy framework; technology and market assessments; strengthening research and development cooperation and trade and investment; and training of renewable energy practitioners
    Our governments should consider a long-term research and development public-private partnership
    Increase collaboration, trade, and investment in specific high-priority areas

• Establish periodic reviews of renewable energy collaboration

NUCLEAR ENERGY

• Collaborate on mutual nuclear priorities: preventing proliferation and theft; properly handling spent fuel and waste; improving safety in design and operation of plants; and improving the economics of nuclear energy
    Simplify and standardize the design of future plants
    Demonstrate long-term storage and disposal of spent fuel and high-level waste

• Our governments and industry should play a leadership role in international organizations relating to nuclear power

• Expand and strengthen bilateral cooperation
    The committee endorses the framework of the Peaceful Use of Nuclear Technology agreement
    The U.S. Department of Energy should consider expanding the Nuclear Energy Research Initiative to involve China and other countries

Consider the model of the National Academy for Nuclear Training to ensure high-quality personnel and safety standards at nuclear facilities

ELECTRICITY TRANSMISSION AND DISTRIBUTION SYSTEMS

- Collaborate on measures to improve the electric power sector in China, including support for international financial institution financing for the grid; the legal limitations of foreign participation in transmission facilities; options to improve adequacy, quality, and reliability of electric power; and reduction of line losses

- Collaborate on developing more economically viable distributed power sources for remote areas and identify cooperative activities that advance distributed generation

1

Energy Setting for China
and the United States

Significant progress has been made in the past 20 years forecasting energy demand, and the increased complexity of modeling systems has resulted in greater accuracy of predictions. Models often are scenario-based and rely on three primary variables: growth in population, economic output, and energy technology characteristics. A rise in either population or economic output results in increased energy demand. Advances in technology and the substitution of more efficient technologies lead to a decrease in energy intensity (the unit of primary energy consumed per unit of economic output), which tempers the rise in energy demand. Availability of capital has a direct influence on how a nation can reduce its energy intensity. It is increasingly the case that the private sector—not the government—provides the capital for energy projects, and it does so by paying greater attention to returns than to societal needs or goals, such as the environment. Projects compete for limited capital on the basis of their potential economic performance: The government role is to shape these markets to best suit national goals.

The issues of natural resource constraints that dominated energy modeling 20 years ago have been superseded largely by the availability of investments in energy resources and infrastructure and by questions of local, regional, and global health and environmental impact from energy production and use. Technology has been responsive to resource constraints; given a market for a particular good or service, an adequate supply generally has been made available. Increased attention to externalities has driven the development of cleaner energy systems through taxation, regulation, and other incentives.

Historical trends have demonstrated a move toward cleaner, more efficient sources and use of energy, an increased share of electricity, and a more diverse mix of primary energy inputs, for the most part using higher quality fuels and processes.

ENERGY TRAJECTORIES FOR CHINA AND THE UNITED STATES

The Committee on Cooperation in the Energy Futures of China and the United States (CCEF) examined existing projections of energy supply and demand through the year 2020. For each country and sector, highlights are presented in a baseline case, followed by alternate possibilities based on possible reshaping of the energy situation, and finally a look at current trends and existing collaborative efforts. A great deal of the U.S. energy baseline case summary is based on the *Annual Energy Outlook* prepared by the Energy Information Administration (EIA) of the U.S. Department of Energy (DOE); much of the Chinese energy sector information is taken from the *China Energy Annual Review*. Chinese data do not include estimates for Hong Kong, Macao, or Taiwan. Other sources are noted in the text. The alternative scenarios also are based on existing projections with sources noted in the text.

A.   ENERGY DEMAND AND END-USE EFFICIENCY

1. U.S. Baseline Case

As shown in Figure 1-1, total energy consumption in the United States is expected to increase by about 1.1 percent per year between 1997 and 2020, from

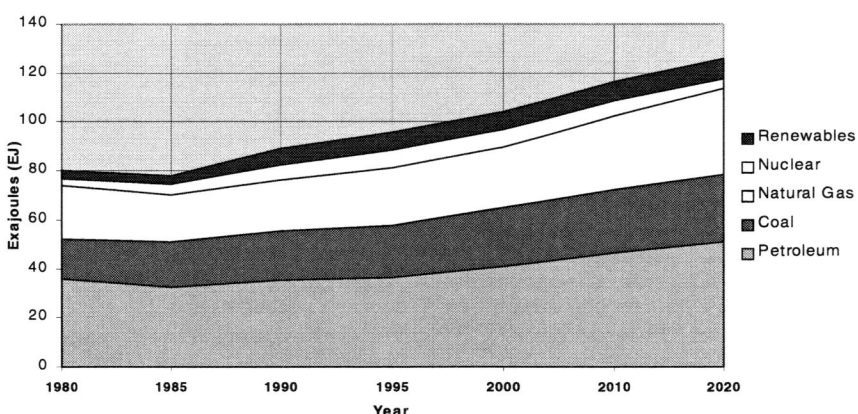

FIGURE 1-1   U.S. Commercial Energy Consumption by Fuel. U.S. Department of Energy, 1999

about 100 to 125 exajoules (EJ).[1] This projection assumes that the United States will have a baseline growth in gross domestic product (GDP) of just over 2 percent per year to 2020, with faster growth occurring in less energy-intensive industries. Energy intensity[2]—the measure of primary energy consumption per dollar of GDP—will continue to decline gradually through 2020.[3] [A number of factors influence the energy intensity—degree of employment of best available end-use technologies, adoption of energy-efficient technologies, as well as structural changes in the economy.] Technological advances and a deregulated, more cost-conscious and efficient electricity sector will result in lower electricity costs in the United States. End-use efficiency improvements will allow increased energy services without significant increases in energy use per capita.

The U.S. transportation sector currently accounts for about two-thirds of total petroleum consumption, which is projected to rise to almost three-quarters in the 2020 time frame. The growth of energy use in the transport sector surpasses the growth of electricity use as the largest single component of increase in energy consumption over this period. Low fuel prices, the absence of new legislation promoting gains in fuel economy, a growing population, and increased travel per capita all contribute to this trend of over 2 percent per year increased demand. Alternative-fueled vehicles—using ethanol, compressed natural gas, liquefied petroleum gas, or electricity—are expected to comprise 8 percent of all vehicle sales in 2020 (about 1.2 million units). Low-emission vehicles are expected to reach 640,000 units by 2020, one-quarter of which will be electric. (EIA,1998).

For commercial and residential buildings, energy use will grow approximately 0.7 percent per year through 2020. The projections are based on continuing relative growth of electricity use in buildings; a substantial increase in the number and intensity of use of plug loads; continuing increases in energy efficiency of space conditioning, appliances, and lighting equipment; and a continuation of current demographic trends (increased population, increased commerce, and larger houses).

Energy demand in the industrial sector is projected to increase by 0.8 percent per year to 2020, with natural gas and electricity being the energy supplies of choice because of ease of handling. Growth in demand for electricity, natural gas, and petroleum by the industrial sector are all expected to rise, by about 30, 26 and 23 percent, respectively. Growth in electricity demand is projected to be met

---

[1] One exajoule (EJ) = 0.948 quadrillion British thermal units (quads). Using higher heating values, 1 billion metric tons of oil equivalent equals 44.9 EJ, 1 billion metric tons coal equal 30.3 EJ, and 1 trillion cubic meters of natural gas equals 39.8 EJ. See energy conversion factors in Appendix.

[2] In calculating energy intensity, electricity contributes to total energy consumption as both end-use consumption and as energy losses.

[3] EIA projects a reduction in energy intensity at an average rate of 1 percent per year from 1997 to 2020, though such projections are difficult to characterize because of the number of variables involved. Other projections are less optimistic about reductions in energy intensity.

by an 80 percent contribution from natural gas power plants and about 10 percent from coal plants.

In all of these sectors the government can play a key role in influencing energy demand by crafting policy and financial incentives to encourage investment in energy-saving technologies to overcome barriers to such investments.

The United States has been a leader in many key technologies and an innovator in some major energy efficiency policies, including auto fuel-economy standards, appliance efficiency standards, and utility demand-side management (from mid-1980s through the 1990s).

2. China Baseline Case

As shown in Figure 1-2 (a scenario developed by the Chinese Academy of Engineering), overall energy consumption in China is projected to grow by more than 50 percent[4] by 2020, from about 45 to 77 EJ. The pattern of end-use energy consumption in China has been relatively stable over the past 20 years. In 1997, industrial energy demand continued to be strong, accounting for about 68 percent of energy end use. Residential and commercial energy demand has resulted in a rapid increase in the use of gaseous fuels and electricity, and the leveling off of coal use. Residential and commercial share of final energy use is about 19 percent.[5] Transport energy demand, although rising quickly, will account for only 10 percent of energy end use by 2020. The agricultural sector has an additional 4 percent share. The U.S. end-use pattern differs dramatically in its much larger share of the transport sector (36 percent) and much smaller share of the industrial sector (36 percent). It is widely believed that China will eventually move closer to this pattern of consumption of the United States and other developed countries (Sinton, 1996).

In recent years China has fallen short of its power capacity expansion plan, but has managed to stay ahead of rapidly increasing energy demand. Power shortages so common a few years ago are for the most part gone in most provinces. This feat can be attributed to significant closures at state-owned enterprises (traditionally large power users) and an impressive record of implementing energy-saving practices.

China has maintained rapid economic growth while sustaining a steady decline in energy intensity. The quadrupling of GDP between 1980 and 1995 was achieved while only doubling the economy's energy demand. This achievement has been associated closely with government policies in economic reform and comprehensive national energy conservation programs initiated in the 1980s.

---

[4] This summary is based on commercial energy consumption and does not include the significant portion of energy from current non-commercial sources in China, estimates of which range from 9 EJ to 11 EJ.

[5] This does not include traditional biomass consumption.

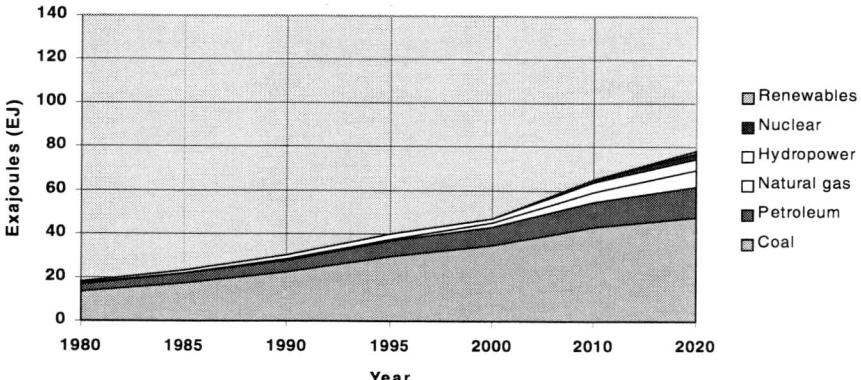

FIGURE 1-2 Chinese Commercial Energy Consumption by Fuel. Chinese Academy of Engineering, 1997

Although substantial success has been achieved, it is widely recognized that significant energy efficiency improvements are still obtainable as major energy-consuming sectors operate at about 15-50 percent less efficiently than OECD countries (Zhou, 1998). Continued commitment to energy conservation is crucial to China's economic and environmental future, and energy efficiency can substantially reduce carbon dioxide emissions from what they otherwise would be.

In numerous applications throughout its energy system, energy conservation can make the most cost-effective contribution to meeting China's energy needs. China is a leading developing country in implementing energy efficiency, and the sum of China's many successes includes reducing energy demand growth to half that of GDP growth since 1980. This is a spectacular achievement that deserves much broader recognition.

Recognizing its rapidly growing transportation sector, China passed the Law of the Highway effective January 1, 1998, as a fuel-based taxation system intended to save energy, reduce pollution and promote automobile technology development.

3. Variations from the Overall Energy and Energy Efficiency Baselines

The Institute for Applied Systems Analysis (IIASA) and the World Energy Council (WEC) in *Global Energy Perspectives* note that scenarios vary in terms of clusters of factors: performance and costs of future technologies; penetration rates of these technologies; availability of energy resources; and geopolitical and policy influences including technology transfer programs, trade, and regulation (WEC/IIASA, 1995). IIASA/WEC also attempts to quantify capital requirements in the energy sector for each of the scenarios and variants proposed. Another group that addressed this difficult issue is the U.S. President's Council of Advisors on Science and Technology (PCAST) Panel on International Cooperation on Energy Research, Development, Demonstration, and Deployment in their *Powerful Partnerships* (PCAST, 1999).

In examining the variation from the baseline case for energy efficiency it is useful to look at energy intensity and technological change to see where the future might differ. Although energy intensity is tied very closely to the state of current technologies, it also includes such factors as structural changes in the economy (e.g., shifts to less energy-intense industries) and changes in energy systems (e.g., shifts from coal to gas).

For the United States, possible variations in end-use consumption include a 25 percent decrease in residential energy consumption (assuming the most efficient technology is chosen, over most cost-effective efficient technology), an 11 percent decrease in commercial energy consumption (energy intensity decreases at 0.6 percent per year), and an acceleration of transportation efficiency that cuts energy use by about 8 percent by 2020.[6]

To provide perspective, in the case of China, it has been estimated that if domestically available advanced technologies and equipment were used to retrofit all backward equipment—much of this in the industrial sector—total energy savings would reach almost one-third of present energy consumption. The question remains, however, what portion of that work can and will be undertaken in the time frame of this study. Capital requirements will be the major determinant of what energy efficiency projects are undertaken, because there is a large stock of inefficient equipment and a long turnover period. Many of these opportunities offer a significant return on investment, though capital remains scarce and capital markets are undeveloped. China Energy Conservation Investment Corporation has recently provided soft money from the government through the State Construction Bank at a level of about $250 million per year (about half for cogeneration projects), though this money will not be available in the future because of fiscal reforms.

In the commercial and residential sectors in China (which are developing quite rapidly, albeit from a much lower position than industry), there is the tremendous potential for the introduction of highly efficient products into a relatively young market. The case in the United States is somewhat different: here the market for such equipment is mature and efficient technologies will make it into the market as replacement goods, not as new stock. Buildings and transportation also both offer considerable energy efficiency opportunities in China.

---

[6] These variations are taken from EIA's alternative scenarios, which some characterize as conservative for two reasons: technology development potential is considered modest, and EIA methodology does not consider the potential impact of new legislation. A great deal of work has been done in characterizing advanced technology development opportunities, especially in the context of emissions reduction scenarios. See in particular the 5 and 11 lab studies and the work undertaken by the Intergovernmental Panel on Climate Change (IPCC).

Traditional fuel use in China—predominantly in rural areas—accounted for 9 EJ in 1996 and is expected to decrease gradually through 2020.

4. Framework for Collaborative Efforts Between China and the United States

The U.S.-China Forum on Environment and Development chaired by Vice-President Gore and Premier Zhu Rongji has built upon the extensive history of interaction between China and the United States[7] to provide a framework to promote cooperation in support of sustainable development. The Forum's goals include the creation of cooperative mechanisms to address local, regional, and global environmental issues, increased deployment of sustainable technologies and practices, and the identification of private-sector opportunities. Four working groups—energy policy, environmental policy, science for sustainable development, and commercial cooperation—meet on a regular basis and have established detailed collaborations in many of the sectors listed below.

The Energy and Environment Cooperation Initiative, signed in October 1997 by U.S. Secretary of Energy Federico Peña and State Planning Commission Vice Chairman Zeng Peiyan,[8] further specifies four priority areas for collaboration between China and the United States: (1) urban air quality; (2) rural electrification and energy sources; (3) clean energy and efficient energy; and (4) Peaceful Use of Nuclear Energy—PUNT, as based on a 1985 bilateral agreement and later agreements.

At the June 1998 Clinton-Jiang summit a number of achievements were announced, including the establishment of the U.S.-China Oil and Gas Industry Forum; commercial contacts for U.S. companies in both coalbed methane and electric power; a financing conference to be held in Beijing; and participation in China's National Air Quality Monitoring Program. This paper will not present a detailed discussion of each initiative, but will mention some of the key agreements and actions undertaken in each sector.

The considerable efforts undertaken on energy efficiency between China and the United States are worth noting. In 1993, the two governments began a process to establish a formal dialogue to share information and promote collaboration on a variety of energy efficiency issues. By 1995, the two countries had reached a formal agreement to pursue mutual objectives in energy efficiency through the establishment of the Sino-U.S. Working Group on Energy Efficiency (under the Energy Efficiency and Renewable Energy Protocol). This working group is composed of representatives of private and public interests in energy

---

[7] Bilateral science and technology cooperation began with China in 1979, and the U.S. Department of Energy currently has over 20 protocols and annexes for cooperation with China on energy efficiency, renewable energy, fossil energy, nuclear energy, and climate change.

[8] Zeng became Minister of the State Development Planning Commission in March 1998.

efficiency. The Working Group has 10 teams, each with co-leaders from China and the United States and a joint membership of 10 to 20 participants. These teams are: (1) energy policy, (2) information exchange and business outreach, (3) district heating, (4) cogeneration, (5) energy efficient buildings, (6) motor systems, (7) industrial process control, (8) lighting, (9) amorphous core transformers, and (10) finance. The Working Group and its teams have remained active and, with limited resources, have been able to sustain joint collaborations in their areas.

There are very substantial multinational efforts in energy efficiency in China—especially the Global Environmental Facility (GEF)—with active public and private U.S. involvement. Some ongoing efforts are focused on creation of a privatized energy management industry, energy-efficient refrigerators, and boiler efficiency, all areas determined to have immediate need and opportunity for improvement.

B.   COAL

1. U.S. Coal Baseline Case

Coal consumption in 1997 was 1,030 million short tons[9] (28 EJ) and in 2020 will reach 1,275 million short tons (35 EJ), an annual growth of about 1 percent. Over 90 percent of coal used in 2020 will be in electricity generation. Coal will remain the largest source of electricity in the United States through 2020, though its share will be down overall. Coal-fired units will account for about half of total electricity generation. Coal-fired units will probably constitute about 9 percent of total capacity expansion (about 32 GW) between now and 2020 (EIA, 1999). There also will be modest increases of about 0.7 percent in overall industrial demand for coal rising to about 80 million short tons (2 EJ) in 2020. Coal use in the residential and commercial sectors will remain constant at about 1 percent of total U.S. coal demand.

Based on experience coal prices will remain competitive through increased mine productivity, and production will increase through 2020. Rail rates for coal are expected to decline in real terms which, coupled with increased fuel efficiency and other increases in productivity, will lead to transportation costs for coal dropping by just over 1 percent per year (EIA, 1999).

2. China Coal Baseline Case

China relies on coal for about 75 percent of its primary commercial energy use. In 1997 raw coal production was about 1,373 million metric tons of raw coal (31 EJ), ranking first in the world and accounting for almost one-third of coal production worldwide. Currently only one-third of China's coal output is used in

---

[9] One short ton (2,000 lbs.) equals 0.907 metric tons.

electric power generation; industry, coke plants, and the commercial sector together account for about 60 percent of coal consumption. Coal production in 2020 will likely be around 1,600 million tons of raw coal (36 EJ).

China recognizes its large dependence on coal use and has adopted numerous measures to limit production, diversify fuel sources, and move to cleaner generation. About 85 percent of China's CO_2 emissions are from coal burning, as are 90 percent of SO_2, 60 percent of NO_x and 70 percent of total suspended particulates (Zhou, 1998).

Most of China's coal resources also are located far from population centers and areas of high energy demand. China's energy infrastructure is best developed around the coal industry, but this system already is stretched to maximum capacity to deliver sufficient energy resources. To demonstrate a level of scale, coal is transported 558 km on average before use. About 50 percent of all freight rail traffic in China is dedicated to coal use, and over 70 percent of coal is moved by rail. Coal also is moved by water and highway, the latter method representing a major new infrastructure capacity. Coal represents about one-quarter of total freight traffic on highways (State Economic and Trade Commission, 1997). China's rapidly increasing energy needs will compound the already excessive strain on transportation systems.

3. Variations from Coal Baselines

There are several scenarios under which coal production and use are affected in the 2020 time frame: The major variance from a reference-case coal trajectory, however, would arise from possible new stringent emission reduction agreements.

For the United States, if a 5.5 percent Renewable Portfolio Standard (RPS) were implemented, new wind, biomass and, to a much lesser extent, geothermal plants could replace coal-fired plants. By 2020 wind generation could reach 52 billion kWh (rather than 8 billion kWh in the reference case), biomass could reach 180 billion kWh, versus 90 billion kWh, and geothermal consumption 34 billion kWh, versus 23 billion kWh (EIA, 1998). In this case coal consumption would decrease by about 100 billion kWh. Carbon emissions are reduced in this case by over 20 million metric tons per year.[10]

In a high energy demand scenario in the United States, however, coal-fired electric generating capacity might be needed to fill a portion of an additional 113 GW of demand growth. Coal could meet about 16 percent of this need under this scenario, and total coal consumption would increase by 12 percent.

---

[10] The emission reduction benefits of a deregulated electric power market incorporating an RPS are impressive: about 60 million metric tons of carbon savings per year. Reductions come from a combination of heat rate improvements, renewable energy, energy efficiency, and distributed generation; increases in emissions come from additional capacity, nuclear retirements, and lower prices (Office of Economic, Electricity and Natural Gas Analysis, 1999).

In China, a program of energy diversification and continued emphasis on higher quality energy resources in the baseline case for 2020 could reduce dependence on coal to about 68 percent of total primary energy consumed. An even more rigorous program of diversification might reduce coal consumption to as low as 56 percent of primary commercial energy used.

4. Current Coal Policies and Collaboration

China is moving to rationalize its coal industry by limiting production from inefficient mines and banning the use of lower quality coal, as well changing the way it uses coal, with a particular effort addressed to lowering consumption of coal in areas in which harmful effects can be minimized or mitigated, such as in electric power generation. Smaller, inefficient coal-fired plants (less than 100 MW) are targets for early retirement to minimize pollution. Mine-mouth plants (so-called coal by wire projects) will take on new importance in the 2020 time frame, as will foreign participation in the coal sector in general. China has been actively pursuing a variety of advanced coal technologies, including coal gasification and liquefaction, coalbed methane production, and coal slurry projects. Approximately 25,000 small coal mines in China, mostly in towns and villages, are scheduled to close in 1999 due to concerns over mine safety and small mine inefficiencies. In the short term, these closures will act as a means of cutting excess production and stockpiles and are also representative of China's desire to rationalize the coal industry over the longer term.

As an abundant and relatively inexpensive energy resource in the United States, coal is used to produce over half of all U.S. electricity. It is, however, a major source of air pollution. As a result of new and proposed air pollution control laws, the cost of building and operating coal-fired power plants is likely to rise relative to other options. Therefore, U.S. policy on coal focuses on research, development, and demonstration of advanced, clean, efficient, and cost-competitive technologies and on the deployment of the technologies domestically and in the major coal markets of China, India, and other countries.

U.S. government support for the development and deployment of clean coal technologies have gone through significant changes over the past 25 years. Initial efforts were aimed at building large-scale demonstration and near-commercial plants using up-front funds provided by the federal government. This strategy failed in part because of the lack of financial incentives for private industry to carry the projects to completion, and in part because of the great uncertainties in the energy marketplace in the late 1970s and early 1980s. More recent programs that call for the government and industry to share the costs and risks of large-scale demonstration projects and for industry to gain the intellectual property from such projects have been more successful. Commercial deployment of some of the clean coal technologies (CCTs) was supported by regulatory (primarily environmental) requirements. However, some of the technologies have not been economically viable in the highly competitive U.S. electric power market, though technologies such as integrated gasification combined-cycle (IGCC) are enjoying market opportunities in other sectors (e.g., oil refining). Regulatory incentives to expedite the initial commercial use of the technologies have been considered but not implemented because they are costly and skew market-based choices.

The following are examples of ongoing cooperative initiatives in the coal sector:

• *The U.S.-China Experts Report on Integrated Gasification Combined-cycle Technology* (DOE and CAS, 1996).
• The United Nations Development Program/Global Environment Facility project and the collaborations between the U.S. Environmental Protection Agency and the China Coalbed Methane Clearinghouse to transfer coalbed methane technology to China.
• The Japanese Green Aid program to provide low-cost SO_2 scrubbers to China.
• Analyses by the World Bank to prioritize CCTs for China.
• Exchanges between the U.S. DOE and Chinese institutions on CCTs (Annex IX of the Fossil Energy Protocol).
• U.S./China Energy and Environmental Technology Center to encourage the responsible development and use of energy in China with an interest in improving the quality of life.

C. NATURAL GAS

1. U.S. Natural Gas Baseline Case

In 1997 natural gas consumption in the United States was 22 trillion cubic feet[11] (25 EJ) and is projected to meet almost 30 percent of U.S. energy demand in 2020, reaching 33 tcf (37 EJ) by 2020, increasing particularly in the industrial sector (a 26 percent rise by 2020). Gas will fill much of the expansion of the electricity capacity (about 88 percent likely will be combined-cycle or combustion turbine technology fueled by gas or oil and gas) and overall will account for a third of total electricity generation capacity in 2020. Additions to domestic U.S. reserves of natural gas could well continue to exceed production through 2020 as a result of increased drilling, higher prices, and productivity gains (Simpson, 1999). The United States has a relatively mature natural gas pipeline infrastructure; however, with natural gas use increasing, more than 75 pipeline expansion projects have been proposed for development over the next few years (EIA, 1998a).

---

[11] One cubic foot roughly equals 0.0283 cubic meters.

2. China Natural Gas Baseline Case

In 1997, natural gas accounted for 2.2 percent of China's primary energy consumption, very little of which was used in power generation. Of the 18.5 billion m^3 (0.7 EJ) that entered the domestic market in 1996, about 40 percent was used as feedstocks for producing fertilizers and chemical fibers, 12 percent was used by households, and most of the rest was used as fuel by manufacturing industries. Natural gas production is likely to be about 25 billion to 30 billion m^3 (1-1.2 EJ) by 2000 and 90 billion m^3 (3.6 EJ) in 2020. If total gas consumption reaches an anticipated 190 billion m^3 (7.5 EJ) in 2020, China will have to increase natural gas imports to 100 billion m^3 (4 EJ) (SETC, 1997; EIA, 1997).

About 3,700 km of major natural gas pipelines are in operation in China, most of which are found in Sichuan Province. A few major gas pipeline projects have recently been completed or are nearing completion, including pipelines to Beijing, Shanghai, Xi'an, and other major urban centers not located in areas rich in gas resources. These are promising signs that the Chinese energy sector is moving toward the use of higher quality, cleaner gas to replace destructive coal burning, particularly in urban areas. Given the potency of methane as a greenhouse gas, an important requirement in the installation of new gas facilities is minimal leakage design.

One intriguing aspect of China's natural gas future is the potential for coalbed methane (CBM)[12] development and use. Having one of the largest endowments of coal resources in the world, China's CBM deposits are believed to be huge: Initial assessments put coalbed methane resources at around 33 trillion m^3 (about 1,300 EJ), comparable to the holdings of natural gas resources (38 trillion m^3 or about 1,500 EJ). In comparison, U.S. CBM resources are estimated to be about 700 tcf (788 EJ), with technically recoverable resources at around 100 tcf (112 EJ) (Gas Research Institute, 1999). U.S. reserves are 11.5 tcf (13 EJ) (EIA, 1998). A preliminary assessment of the regional distribution of China's CBM resources is found in Table 1-1. Coal mining activities in China currently release about 19 billion m^3 (0.75 EJ) of CBM per year, a much more potent greenhouse gas than carbon dioxide.[13] Currently coal mines withdraw about 500 million m^3 (0.02 EJ) for local use.

3. Variations from Natural Gas Baseline Cases

In a higher electricity demand scenario in the United States, natural gas use could increase by 17 percent from the baseline case, meeting over 80 percent of additional required electricity capacity.

---

[12] The process that converts plant material to coal produces large quantities of methane-rich gas, which is stored within the coal. This coalbed methane has only recently been treated as an energy resource, though its utilization poses some new technical difficulties.

[13] Coalbed methane is a greenhouse gas with a potency of about 50 times that of CO_2 in a 20-year time frame and about 20 times as potent on a 100-year horizon.

TABLE 1-1 Regional Distribution of CBM Resources in China

Area	Resources (billion m³)	Percentage (%)
Northeast	2,479.8	7.6
North China	21,125.4	61.7
Northwest	5,063.6	15.5
South	4,967.8	15.2
TOTAL	32,626.6	100.0

Chinese natural gas production and use could be increased significantly through the adoption of aggressive measures to develop this new domestic industry. For example, in *China's Least Cost Power Options*, the possibility was presented that natural gas could play a much more significant role in meeting China's power needs. In this analysis, electric power from natural gas could meet up to one-third China's future power needs at a lower total cost than using coal—given the domestic production of low-cost gas and local manufacturing of gas turbines (Battelle Memorial Institute, 1998). This scenario would entail dramatic policy reform and significant transfer of technology in both gas production and power generation.

Development of gas resources and infrastructure would play a critical role for much greater expansion beyond 2020. The three new long–distance natural gas pipeline projects from the Ordos Basin to Beijing, Xi'an, and Yinchuan are a good example of early efforts to increase marketability of gas. Additionally, exploitation of Chinese CBM resources could greatly speed the development of the natural gas industry and significantly reduce greenhouse gas emissions.

4. Current Natural Gas Policies and Collaboration

In March 1998 the National People's Congress approved reform measures that, among other things, allowed for the first time the integration of upstream and downstream activities in independent national corporations, each with a regional base of operations. Since then, China's oil and gas industry has seen the largest restructuring in its history. The full ramifications of this move remain to be seen, but it is hoped that the restructuring will improve productivity, increase competition, and ultimately reenergize China's oil and gas sector to meet future demand. China also took a major step in oil price reform in June 1998 by pegging domestic oil prices to international standards, moving further ahead in integrating domestic oil industry with that of the world.

The China United Coalbed Methane Corporation (CUCBM) established in 1996 is solely responsible for the exploration and development of CBM, and for foreign participation in this field. The government plans to raise the output of CBM from 0.5 billion m³ in 1997 to 1 billion m³ (0.04 EJ) by 2000, and to 10 billion m^3 (0.4 EJ) by 2010. In comparison, the United States currently produces about 1.1 trillion cubic feet (1.2 EJ) of CBM annually, about 6 percent of total U.S. natural gas production. CBM reserves in the United States are estimated at about 11.5 tcf (13 EJ) (EIA, 1998). Box 1-1 presents opportunities for potential collaboration in the CBM industry.

Liquefied natural gas (LNG) is authorized for trial in Guandong Province: this represents a significant opportunity for sale of equipment, technology, and services necessary to build and operate an LNG facility, as well as the long-term import of LNG from Alaska to China.

The U.S.-China Oil and Gas Industry Forum works to increase private-sector participation in China's energy development. Key elements of the agenda of this public-private partnership group include issues of mutual access to information for firms interested in either country's markets; natural gas and oil policies that build on work started in the Asia Pacific Economic Cooperation forum (APEC) natural gas initiative, to accelerate investment in natural gas and oil that support supply, transportation, and market development in China; cooperative opportunities in third-country projects, including exploration, development, transportation,

---

BOX 1-1   Potential Collaboration in the CBM Industry [a]

- Research on the process of CBM accumulation
- Development of cost-effective tools and methods for exploration and evaluation of CBM resources
- Prediction techniques for the CBM-enriched regions with high permeability
- Classification of and calculation methods for CBM reserves
- Research on predicting the output of CBM and optimizing a development program
- Techniques for drilling, completion and simulation of CBM with complicated coal geological features
- Fracturing techniques and coal reservoir protection for CBM wells with complicated coal geological features
- Techniques for enhanced CBM primary recovery
- A study of the mechanism and measurement of coal seam permeability
- An economic evaluation of methods and management of CBM production
- An evaluation of environmental implications of CBM development and production.

[a] Many of these technologies are proprietary and details of collaboration would have to be developed in a commercial context.

---

and refining; and business practices that explore conditions particular to the oil and gas industries that need to be addressed in an international context. The EPA is also working with the State Development Planning Commission to collaborate on natural gas policy reform.

With funding from the GEF, UNDP has helped to transfer technology to Chinese companies involved in coalbed exploration drilling in an attempt to better prepare local industry for the challenges of this new industry. Additionally, a number of private firms have signed CBM contracts with the CUCBM and testing is underway. The United States and China are actively collaborating on possibilities in CBM through the Oil and Gas Industry Forum as well as through a number of initiatives through DOE, EPA, and CUCBM.

D. PETROLEUM

1. U.S. Petroleum Baseline Case

In 1997, U.S. petroleum consumption was about 39 EJ (37 quads) and probably will continue to rise steadily through 2020 with total consumption at about 50 EJ (48 quads). The petroleum share of total energy consumption remains fairly steady at around 40 percent. Transportation and industrial demand account for much of this projected growth, with slight declines in residential, commercial, and electric utility use. The price of crude oil is expected to rise moderately and decreasing exploration costs could prompt increased drilling.

U.S. crude production continues to decline by about 1 percent per year through 2020, a decline from about 15 EJ (14 quads) to 11 EJ (10.5 quads). Enhanced oil recovery follows petroleum prices very closely and thus only happens at higher oil prices.

Oil imports make up the difference in declining domestic production and increasing consumption. In 1997 the United States imported about half of the petroleum that it consumed; in 2020, that share will be closer to two-thirds. Total oil imports in 1997 were about 10 million barrels per day (mbd) and are projected to increase to over 16 mbd in 2020. Share of OPEC imports from Persian Gulf nations rises from 39 percent in 1997 to almost 50 percent in 2020 (EIA, 1999). Imports of refined petroleum products also will rise, from 11 percent in 1997 to 20 or 30 percent in 2020 as demand outpaces the only modest increases in domestic refining capacity.

2. China Petroleum Baseline Case

Oil represents less than 17 percent of China's total primary energy consumption,[14] and China's crude oil production capacity of 160.7 million tons (7.1 EJ) in

---

[14] This includes all forms of commercial energy consumption, including renewables, but does not include traditional energy use.

1997 ranks fifth in the world. Nonetheless, China became a net oil importer in 1993 and a net crude oil importer in 1996 and by 2000 likely will import 50 million tons (2.1 EJ) of crude and oil products. By 2020 China is projected to produce about 220 million tons (9.2 EJ) of oil and import about 130 million tons (6 EJ).

Over 19,000 km of pipeline systems (about 70 percent of which carry crude oil) are in place in eastern China, mostly along the coast or on rivers near China's refineries. Oil fields in western China are not yet connected via pipeline, instead relying on railway systems. Refined oil products are currently moved by rail (60 percent), water (30 percent), and pipeline (10 percent).

3. Variations from Petroleum Baseline Cases

For the United States, EIA has developed alternative case comparisons for high and low world oil prices, each about one-third higher and lower respectively than the reference case in 2020. Domestic oil production and consumption patterns are linked closely to price fluctuations and the economy reacts quite rapidly. U.S. petroleum consumption ranges from 46 EJ (44 quads) in a low-growth/high oil price case to almost 56 EJ (almost 53 quads) in a high-growth/low oil price case. In the high-technology case, total fuel consumption for the transportation sector is about 8 percent lower than the reference case level in 2020. Assuming successful technology development and subsequent marketplace adoption, the transport sector could reduce carbon emission by 100 million to 180 million metric tons per year (U.S. Department of Energy, 1996).

China would very much like to decrease its dependence on imports of petroleum and petroleum products. The possible options range from increasing domestic production, decreasing dependence on petroleum, and development of alternative liquid fuels: all of which hinge on major increases in investment and deployment of technology. The refining industry in China also will play a significant role in alternative trajectories as demand increases for higher quality products—transportation fuels with reduced lead and sulfur—while quality of crude oil decreases. Other estimates of China's oil import balance are more severe: EIA projects that China's petroleum capacity in 2020 will only deliver less than 4 mbd of the over 10 mbd estimated to be consumed in 2020.

4. Current Petroleum Policies and Collaboration

The U.S.-China Oil and Gas Industry Forum addresses issues of oil and gas policy, technologies, business practices, and trading in an effort to assist China in securing "reliable, economic, clean, and abundant" sources of oil and natural gas (U.S.-China Oil and Gas Industry Forum, 1997). See section on natural gas, above, for highlights of this agreement.

Through development of the petroleum industry over the last 40 years, China has a relatively complete portfolio of science and technology available, many of which are at world levels. Much of this science and technology has been developed around the on-shore oil industry which only represents a fraction of China's total reserves. Box 1-2 contains a list of some of the exploration, development, and drilling technologies which could benefit from increased technology cooperation.

E. NUCLEAR POWER

1. U.S. Nuclear Power Baseline Case

With the current deregulation of the electric power sector in the United States, there are many questions concerning the 105 nuclear power plants currently operating. It is anticipated that during the time frame covered by this study the total nuclear capacity in the United States will decline from 99 GW in 1997, representing 18 percent of total electricity generation, to 50 GW in 2020, 7 percent of total electricity generation (EIA, 1998). This corresponds to the retirement of 65 nuclear units during this period, and in 2003 natural gas is expected to surpass nuclear as the nation's second largest source of electricity.

Though a significant number of older plants will be retired, the Nuclear Regulatory Commission has defined an application process to extend a plant's operating license for an additional 20 years. In 1998 two U.S. utilities chose to submit license renewal applications on the assumption that capital costs associated with extending the life of an existing plant would be lower than building new electric power generation capacity; others are considering applications.

Nuclear plant performance continued to improve with 90 percent of plants having production costs competitive with coal and natural gas. Capacity factors continue to improve as outage management for refueling and maintenance is shortened—national load capacity averages are expected to rise from 75 percent in 1996 to 85 percent in 2020. All plants typically run continuously between refuelings and, given their low fuel costs, will continue to do so, even in a deregulated market. Safety priorities are of the highest concern, and plants will be taken off line occasionally between outages for inspection purposes.

The United States has opted not to reprocess spent fuel for its future fuel value. Rather, it is evaluating a site for permanent disposal of its used fuel. This decision remains controversial and the presumption of permanent disposal might be reviewed before permanent burial of fuel is initiated. Some nuclear power nations view spent fuel as a source of fuel that would make fission essentially sustainable from a resource point of view. However, current economics do not justify reprocessing. The abundance of natural uranium is much greater than had been predicted, and some argue that unlimited amounts of uranium are available from seawater at prices competitive with the extra costs of fabrication of mixed-oxide (MOX) fuel.

BOX 1-2   Chinese Technology Priorities in the Petroleum Industry

Exploration
- Exploration technologies in Craton Paleozoic carbonate sedimentary basins and clastic rock sedimentary basins of foreland basins
- Prediction technologies for oil and gas in shoal, lithologic trap and deep Paleozoic strata under 5,000 m
- Natural gas theory and comprehensive matching exploration technologies on Paleozoic marine facies strata and meso-Cenozoic terrestrial formation
- High-maturity oil and source-rock correlation in marine facies strata of Tarim Basin
- Determination of maturity of high- and low-maturity source rock
- Research and application of terrestrial sequence stratigraphy
- Research on hydrocarbon migration
- Research and application on four-dimensional seismic imaging
- Dynamics study on hydrocarbon accumulation
- Basin dynamics theory
- Seismic exploration technique for complex ground features and high dip angle structures
- Prediction on subtle trap and research on high-resolution sequence stratigraphy
- Hydrocarbon detection and reservoir prediction technologies using seismic data

Development: reservoir characterization, mapping, and enhanced recovery techniques
- Gas injection (CO_2, N_2) enhanced recovery techniques; particularly for recoverable reserves in late high water-cut period in the eastern oil fields with water injection development

---

2. China Nuclear Power Baseline Case

China's nuclear power program began in the 1980s with their first commercial plant Qinshan, a 300-MW pressurized water reactor (PWR) going online in 1991. In 1994, Daya Bay (two French 900-MW PWRs) went online providing 2.1 GW of nuclear capacity, 1.3 percent of total electric power generation. An additional 6.4 GW of capacity is under construction, including Qinshan phase two (two Chinese-design 600-MW PWRs) expected to go online in 2002 and 2003, Daya Bay phase two (two additional French 900-MW units) in 2002 and 2003, Qinshan phase three (two Canadian 700-MW (CANDU reactors) expected

- Decreasing costs of enhanced oil recovery (EOR) techniques in tertiary recovery
- Matching technologies for cost-effective and economic development of low and extremely low permeability oil fields
- Development technologies for special hydrocarbon accumulations
- Technologies of steam injection and ultra-heavy oil development for heavy crude oil
- Natural gas injection exploitation methods for gas condensate reservoirs
- Geological modeling technology and reservoir simulation, and commercialization of reservoir numerical simulation software
- Research on the distribution of remaining oil and potential exploitation technology
- Reservoir protection technology
- Interpretation technology of test wells
- Problem of high-temperature resistance of water shutoff agent

Advanced Drilling Engineering
- Unbalanced drilling and completion techniques for low-permeability hydrocarbon reservoirs
- Multilateral drilling techniques
- Long-distance reach drilling techniques
- Slim-hole drilling techniques
- Drilling techniques for high-temperature and high-pressure conditions and for ultra-deep wells

to go online in 2003, and Tian Wan (Lianyungang, two Russian 1000-MW PWRs) expected to go online in 2004 and 2005.

Although small by comparison with other developed nuclear programs, this program represents a significant commitment to nuclear power as an important component of China's future energy mix. Nuclear power plays a strategic role for the densely populated coastal areas. China is aggressively developing its own 1,000-MW PWR design to serve as the backbone of this program early in the next century, and hopes to increase nuclear capacity to 20 GW by 2010 and 40 GW by 2020, though capital constraints could prove daunting.

China's policy is to reprocess its spent fuel and a pilot processing facility is under construction. China intends to recycle plutonium as MOX fuel in PWRs or in breeder reactors. China has elected to construct four regional low- and intermediate-level waste facilities and is contemplating vitrification and geologic isolation of high-level waste. China's National Nuclear Safety Administration (NNSA) regulates nuclear safety in concert with the International Atomic Energy Agency (IAEA). China is also a signatory to the Nuclear Non-Proliferation Treaty. China is expected to play an active role in the World Association of Nuclear Operators (WANO), which will accelerate the ability to safely operate the different reactor systems under construction.

3. Variations from Nuclear Baseline Cases

Nuclear power could play a more significant role in a high economic growth or a high world oil price case for the United States. In such cases it is possible that nuclear power in 2020 could increase slightly to contribute about 360-375 (of a total 4,450 to 4,800) billion kWh.

The U.S. interest in nuclear power could also increase if CO_2 controls are imposed and nuclear power could contribute to reduction of NO_x and SO_2 emissions under the Clean Air Act. Existing plants would be more highly valued, and more plants would likely seek 20-year license extensions. Interest would likely increase in consolidating ownership to lower operating cost, and new orders for U.S. plants could occur before 2020. Under a carbon constraint scenario, existing nuclear plants could gain immediate economic benefit from selling CO_2 credits to coal plants that would have to buy credits or control CO_2 emissions.

Chinese nuclear plants could also benefit in the establishment of CO_2 markets with credits to offset either foreign or domestic CO_2 releases. These credits could help meet the capital requirements necessary to realize China's ambitious nuclear power goals.

Outside of the time frame of this study, potential variations from the baseline case vary widely. For example, the ecologically driven scenarios developed by the IIASA/WEC team put nuclear power contribution at a crossroads: in one variant nuclear is assumed to be a transient power source and will ultimately be phased out (though well beyond the limit of this study); in the other variant a new generation of safe, modular reactors (150 to 300 MW electric) fills the electricity needs of densely populated urban areas where conversion of renewable energy to electricity is impractical. The latter variant requires significant reduction in capital costs and public acceptance of nuclear power—the conditions for which are discussed elsewhere in this paper.

4. Current Nuclear Policies and Collaboration

The Agreement on Intent of Cooperation Concerning Peaceful Uses of Nuclear Technology (PUNT), signed by the United States and China in October

1997, established the framework to, among other things, exchange technical information—including joint research and development (R&D) projects—on current and advanced light water reactor technologies; improve plant design, safety, and economic performance; address fuel and waste treatment and storage; and technology develop to enhance international nuclear safeguards. U.S. policy now permits U.S. companies to sell commercial nuclear power technology in China.

The U.S. Department of Energy's Office of Nuclear Energy, Science and Technology is emphasizing a Nuclear Energy Research Initiative (NERI) in response to recent recommendations made by PCAST. The basic premise is that if issues of cost, safety, proliferation risk, and waste disposal can be resolved, nuclear energy can make a positive contribution to the U.S. energy future: by addressing carbon emissions and other environmental impacts associated with energy production; by decreasing reliance on imported energy sources; and by increasing exports of energy technologies. The initiative will address proliferation-resistant fuel cycles, new reactor designs, advanced nuclear fuels, new technologies for waste management, and fundamental nuclear science. A peer-review selection process will consider innovative proposals from universities, national laboratories, and industry, based on their technical excellence and relevance to the following long-term strategic needs: developing new reactor and fuel concepts, maintaining U.S. leadership in nuclear technologies, and promoting and maintaining nuclear science to help meet future challenges.

The Institute of Nuclear Power Operations (INPO) oversees operator training and performance to ensure that best practices are communicated among all parties. Periodic plant ratings provide an indicator of plant performance to owners and the public and contribute to the Nuclear Regulatory Commission's independent assessments. WANO provides international oversight and provides a means of sharing good practices. The IAEA provides an independent assessment of national nuclear power programs with respect to safety. It also has produced guidelines with regard to security of fissionable material. Given the high profile of any nuclear incident and its effect on public confidence in nuclear power, the role of the IAEA is important and well supported by the international community.

The Chinese National Nuclear Safety Administration (NNSA) and the U.S. Nuclear Regulatory Commission began collaborating in 1984. This has led to the mutual exchange of expert individuals for training, consulting, and exchange of experiences. These activities augment U.S. and Chinese participation in WANO and the IAEA.

The nuclear energy industry in the United States and Japan also has been actively collaborating with Chinese institutions to design plants suitable for the Chinese market. The European advanced PWR also has potential for the Chinese market, given the early Chinese commitment to PWR technology.

F. ELECTRICITY

1. U.S. Electricity Baseline Case

In 1998, total electricity generating capacity in the United States was 778 GW, producing 3,620 billion kWh of electricity. In 1998 coal-fired plants accounted for about 52 percent of total electricity generation, nuclear 18 percent, natural gas 15 percent, hydro 9 percent, petroleum 4 percent, and renewables about 2 percent.

In the United States, electricity demand is projected to be 4,345 billion kWh by 2020. This will require a total of 363 GW of new capacity, of which 126 GW will replace retired units and 237 GW will reflect demand growth. Residential and industrial electricity demand will rise; commercial demand also will rise, but efficient equipment, particularly lighting, motors, heating, cooling, industrial processes, and building materials will temper increases in demand in this sector.

The cost of electric power generation through 2020 probably will decrease by almost 1 percent per year, resulting in projected prices for residential, commercial, and industrial customers being 15 to 20 percent lower than in 1997.[15] In the case of coal-fired plants, steadily declining fuel costs have reduced generating costs by almost half over the period 1980 through 1996. Fuel prices for natural gas have increased and are expected to continue to do so, though additional efficiency gains have offset these costs, lowering generating costs by 25 percent from their peak in 1984.

2. China Electricity Baseline Case

China's electricity generation capacity is second only to that of the United States, reaching 250 GW in 1997 and total electrical production reached 1,081 TWh. In the short-term China is experiencing an excess in capacity and has taken this opportunity to close many small, inefficient, and environmentally damaging thermal power stations. Major power construction projects have been deferred for three years. Over the next half century, China is expected to continue large-scale expansion in electric power to meet targets of modernization. According to original planning, the targeted installed capacity is scheduled to reach 290 GW by the year 2000, 500 GW for the year 2010 (of which hydropower accounts for 115 GW, nuclear power contributes 20 GW), and probably 700 GW for the year 2020. Recent economic difficulties as a result of the Asian economic crisis, limitations on capital availability, and other factors, however, will slow this ambitious plan of power capacity expansion.

About 14 percent of exploitable hydro resources have been developed in

---

[15] An increase in retail competition for electric power is expected to contribute to lower electricity costs, partially through efforts to streamline utility operations and partially from deployment of advanced technologies to lower operating costs.

China, and in 1998 hydropower accounted for over 8.4 TWh of increased generation. The Three Gorges Dam is the largest hydropower project under construction; when completed it will add 18 GW of electric power capacity. Some of the significant obstacles to increased development of hydropower are high capital costs, long payback periods, inaccessibility (i.e., distance from population centers and energy demand) of hydro resources, and site-specific concerns over ecological consequences.

Household electricity consumption in China is 330 kWh per household per year, or about 10 percent of U.S. consumption. More than 860 million Chinese (about 70 percent of the population) live in rural areas with inadequate access to commercial energy, and, among them, in 1998 about 40 million have no access at all to electricity (Chinese State Power Corporation, 1999). This segment of the population depends heavily on biomass energy and suffers from the consequent negative health and environmental impacts from burning of these traditional fuels. The Chinese government is making a large investment to provide electric services in rural areas to alleviate this problem.

One impediment to this objective of providing commercial energy to rural areas is that the development of the power transmission and distribution (T&D) network is far from complete. Compared with power grid systems in developed countries where meshed networks have been formed, the power networks in China (six regional networks and several independent grids) are still in the early stages of development. The framework of the system is not strong enough in either system security or capacity. It is necessary to strengthen the T&D system by constructing new lines as well as upgrading old ones.

Growth in the urban electric power distribution network in the 1990s was between 14 and 18 percent per year, reflecting China's urbanization, unprecedented in scale. Overloading of the network remains the bottleneck of meeting reliable power supply: updating and upgrading this system is an urgent need. The Chinese government is currently undertaking a program to upgrade the urban distribution system with an investment of about 200 billion RMB (about U.S. $24 billion).

Electricity transmission line losses were reduced from almost 9 percent in 1981 to 8 percent in 1990, though this figure does not include losses in low-voltage networks and in the supply networks owned by large consumers which could be as high as an additional 7 to 8 percent. Taking into account power plant energy use, only about 75 percent of generated electric power reaches the end users in the worst case. The distribution network suffers from the same problems, especially in rural areas where distribution line losses, including nontechnical losses, can be as high as 25 percent.[16]

---

[16] In the United States, energy used at the plant is about 5 percent and line losses generally range from 5 to 8 percent.

3. Variations from Electricity Baseline Cases

The trend of increased use of higher quality energy through electrification is one that pervades all energy trajectories; consumers are increasingly demanding cleaner and more convenient forms of final energy. Another trend is the increased efficiency of combustion, with goals of having coal-fired electricity attain 45 percent efficiency, and gas reaching about 60 percent.

High economic growth projections for the United States[17] could lead to an increased electricity consumption of about 350 billion kWh in 2020. For both the United States and China decentralized markets also could play a larger role in alternative trajectories, especially using renewables (addressed in the next section) and natural gas and CBM (considered above).

Use of petroleum as a utility fuel likely will disappear almost entirely (from an already insignificant level). Overall, an optimistic trajectory for introduction of electricity in China could envisage meeting the goal of 700 GW of installed capacity by 2020, combined with major progress in extension and quality of transmission and distribution systems.

4. Current Electricity Policies and Collaboration

The current trend in electric power markets in both China and the United States is toward a deregulated competitive system. Though the two countries are at different stages in their restructuring, both intend to separate generation facilities from transmission and distribution networks. This situation presents an important opportunity for collaboration between our two countries. There is also the need to coordinate closely with international financial institutions who are providing support to China in efforts to create a competitive generation market, one that includes independent power producers. Additional discussion of deregulated energy systems is found in Chapter 2.

As a major feature of the *Energy and Environment Cooperation Initiative*, electricity supply, especially for rural populations, is a high-priority item for U.S.-Chinese collaboration. Many of the DOE agreements and protocols in place also address development of cleaner electric power through both fossil fuels and renewable sources.

The Federal Energy Regulatory Commission has also conducted exchanges and activities with Chinese counterparts, as has the Electric Power Research Institute (EPRI). China currently is participating in EPRI's R&D and services programs.

---

[17] In the Energy Information Administration high economic growth case, GDP increases at an annual rate of 2.6 percent compared to 2.1 percent growth in the reference case.

G. RENEWABLE ENERGY

1. United States Renewable Energy Baseline Case

In the United States, nonhydropower renewable energy is expected to contribute 3 percent of total electric capacity by 2020, up from 2 percent in 1997. Rapid growth is expected by 2020 in biomass (20 percent increase to 90 billion kWh), geothermal (an increase of nearly 50 percent), and municipal solid waste (increased to 30 billion kWh). Wind electric generating capacity in the United States declined in the mid 1990s but a resurgence of new construction since 1998 is expected to add about 1 GW by 2000 to the 1.88 GW of 1997 grid-connected wind power capacity. In the 2020 time frame, growth is expected to continue at about 3 percent per year to contribute a total of about 3.6 GW by 2020.

Solar energy consumption in the United States remained flat in the past five years because of a lack of new installations in a market undergoing deregulation (EPRI/DOE, 1997) However, shipments of solar photovoltaic (PV) cells and modules and solar thermal collectors have increased sharply, largely because of a strong export market. Neither solar thermal nor solar PV is expected to make a large contribution to grid-connected electricity supply by 2020, though their application in niche markets will continue to grow.

Conventional hydropower, which supplied about 10 percent of U.S. electricity in 1997 (360 billion kWh), is expected to gradually reduce its share to about 7 percent in 2020 (330 billion kWh). Because of the reduced share of hydropower, when grouping all renewable energy sources (including conventional hydro) overall share of electricity supply from renewable sources will probably drop from 12 percent in 1997 to 10 percent in 2020. See Table 1-2 for renewable energy generation predictions.

TABLE 1-2 U.S. Baseline Case Renewable Energy Generating Capacity (thousand MW)

Energy Source	1997	2020
Conventional hydro	77	78
Geothermal	3	3.5
Municipal solid waste	3.4	4.3
Wood, other biomass	1.7	5.6
Solar thermal	0.4	0.5
Solar photovoltaic	0.01	0.6
Wind	1.9	3.6
TOTAL	88	97

2. China Renewable Energy Baseline Case

Since the 1970s the Chinese government has recognized the importance of active development and application of renewable energy for off-grid rural and remote areas, and this work has been included in the national five-year plans. Through a continuous effort over 20 years, significant progress has been achieved. However, renewable energy resources are not anticipated to make a significant contribution to on-grid electricity capacity by 2020. Table 1-3 shows the current status of renewable energy deployment in China. Table 1-4 shows another set of renewable energy development projections for China to 2020.

In comparison to the United States, China's wind power development is small but growing rapidly. In 1998 grid-connected capacity was 240 MW, compared to 167 MW installed capacity in 1997 and 57 MW in 1996. There are also about 17 MW of off-grid small wind units in operation. China has world-class wind resources with an estimated technical potential of 250 GW, although much of it is far from population centers. Initial site assessment has identified 3-8 GW. Further development of wind power resources will be dependent on advances in energy storage or backup systems to account for the inherent intermittent nature of this resource.

China has the world's largest and fastest-growing market for solar hot water heating. Over 5 million m^2 (heat-absorbing area) of solar water heaters had been

TABLE 1-3   Development Status of Renewable Energy in China (Institute of Electrical Engineering, Chinese Academy of Sciences)

Energy	Item	Present Situation
Biomass	Biomass digesters	About 5.25 million sets, 1.47 10^9 m^3/year
	Firewood forest	About 5.4 million hectares
Mini-hydro	Power stations	>60,000 stations, about 17,000 MW
		34.3 billion kWh
Tidal	Power station	8 stations, 11MW
Geothermal	Power stations	5 stations, 28.78 MW
	Direct use	1.6981 x 10^4 TJ per year
Wind	Mini-Generators	150,000 sets, 15 MW
	Water lifting machines	>2000 sets, 2.11 MW
	Wind farms	19 farms, 167 MW
Solar	PV cells	~8.8 MW
	Hot water heaters	~5 million m^2
	Solar houses (passive)	2.7 million m^2
	Greenhouses	0.342 million hectares
	Dryers	20,000 m^2
	Cookers	150,000 sets

TABLE 1-4 Projections for Renewable Energy Development in China (Institute of Electrical Engineering, Chinese Academy of Sciences)

Source	1990	2000	2010	2020
Solar Thermal Utilization				
Water heater				
Mm^2	1.5	9.0	15.0	30.0
Solar house				
Mm^2	0.4	10.0	20.0	100.0
Thermal Power Generation				
GW	—	—	0.1	2.0
TWh	—	—	0.2	4.0
PV Power Generation				
GW	0.002	0.015	0.3	3.0
TWh	—	0.05	0.9	7.0
Wind Power Generation				
GW	0.02	0.35	1.1	6.0
TWh	0.05	0.95	3.0	17.4
Geothermal Utilization				
mtce	0.35	0.8	2.0	6.0
Power Generation				
TWh	0.1	0.3	0.5	1.0
Biomass energy				
mtce	263	240	260	290
Traditional Technology	262	236	240	200
New Technology	1	4	20	90
Power Generation				
GW	—	0.05	0.3	3
TWh	—	0.2	1.2	12
Ocean Energy				
GW	0.01	0.05	0.6	5
TWh	—	0.1	1.6	15
TOTAL				
mtce	264	242	296	315
Power Generation				
GW	0.04	0.5	2.6	20.0
TWh	0.15	1.6	7.6	60.4

installed as of 1996. Other solar thermal applications include passive solar-heating houses and solar cookers. The PV market in China is small but growing quickly, with about 8.8 megawatts-peak (MWp) power in 1996. About 50 percent of existing PV power is used for telecommunications, 10 percent is used for industries, and most of the rest supplies electricity for remote areas without grid coverage. Solar thermal power generation is still in the R&D stage.

China has made great efforts to improve the efficiency and technology of biomass utilization through national programs for efficient stoves and rural household biogas digesters and, more recently, commercial applications in bagasse co-generation and biogas power generation. Total biomass consumption amounted to about 9 EJ in 1996, over 90 percent of it firewood and crop stalks burned for household energy needs. The United States consumes about half as much biomass fuel, but in more modern applications. Among the about 3 EJ consumed in the United States in 1996, 75 percent was used in industry, and 22 percent was consumed by households for space heating. About 27 percent of the U.S. biomass fuels are used for electricity generation.

3. Variations from Renewable Energy Baseline Cases

Electric power from renewable energy sources often is envisioned as the long-term goal of a nation's energy development, though the time frame in which such a transition can occur remains unclear. The major factors influencing the widespread use of renewable energy resources are price and policy measures to better balance energy needs with local and regional environmental objectives and the framework of a national and global regime to address emissions of greenhouse gases.

Even in an accelerated worldwide renewable energy deployment scenario, the contribution to be made by renewable sources is more gradual in the near to midterm (WEC/IIASA, 1995). Under an aggressive system of policy and financial incentives as well as significant technology transfer efforts and international cooperation, renewable energy can make a large contribution beyond the time frame of this study (perhaps by 2050), but in order to do so, R&D, investment, financial incentives, and collaboration on policies and technologies must begin now.

There is a possibility in the United States, through an aggressive program of incentives and R&D, of almost doubling non-hydro renewable energy contribution to electricity capacity to 6 percent by 2020, adding, in particular, over 20 GW of wind power rather than the 3.6 GW in the reference case. Greenhouse gas emission reductions under this scenario would be significant: a 70-million-ton reduction, or 3.5 percent.

New programs to boost solar energy use are being implemented by the United States. DOE launched the Million Solar Roofs Program in 1997, and local projects for solar PVs are being launched.

China's New and Renewable Energy Development Outline from 1996 to 2010 requires that commercial renewable energy consumption increase from the current level of less than 2 mtce to about 120 mtce by 2020.[18] Biomass comprises the bulk of the energy provided (about 90 mtce), followed by solar thermal appli-

---

[18] This figure does not include current noncommercial use of renewable energy. According to the IEE projection, traditional biomass energy use will decrease to about 200 mtce by 2020.

cations (totaling 8.5 mtce), then wind energy (6 mtce), ocean energy (5 mtce), and solar PV and geothermal (each at 2.5 mtce).

4. Current Renewable Energy Policies and Collaboration

The Chinese State Science and Technology Commission-DOE 1995 Protocol on Energy Efficiency and Renewable Energy and its six annexes provide a number of opportunities for collaboration in solar, biomass, wind, hybrid systems, geothermal, electric vehicles, and so on. In 1998, U.S. funding for the program increased to about $1 million, compared to $400,000-$600,000 in previous years.

As part of the Energy and Environment Cooperation Initiative noted earlier, part of a now $100 million loan program at the U.S. Export-Import Bank has been earmarked for U.S. companies interested in developing renewable energy projects in China (other projects eligible under this program include energy efficiency and small-scale clean coal projects).[19]

Other ongoing efforts in renewable energy include:

• World Bank/Global Environment Facility demonstration of wind (~290 MW) and solar power systems, under way since 1996 and funded at over $400 million;
• U.S. DOE-Chinese Ministry of Agriculture: biomass cooperation under way since 1996;
• Energy Research Institute-National Renewable Energy Laboratory collaboration: under the Renewable Energy Protocol, Annex 1 (Center for Renewable Energy Development);
• UNDP/GEF renewable energy project of over $25 million half the funding of which is designed for pilot projects in biogas, bagasse, and hybrid village power, the other half of which is for technical assistance, training, codes, standards, resource assessment in solar, wind, geothermal, and biomass; and
• Asian Development Bank funding prefeasibilty studies and assessments in biogas and wind at about $1 million.

In addition to ongoing collaboration with the United States, China also is receiving bilateral assistance in renewable energy technologies from Denmark, Holland, Germany, Spain, and Japan. Australia and Spain are also co-donors to the UNDP project.

---

[19] The committee recognizes the controversy surrounding this initiative and later in this report offers suggestions on how to make better use of this facility.

2

Perspectives and Commentary

China and the United States share four important challenges in achieving their energy goals. As very large economies, each lacking adequate domestic oil reserves, we share a dependence on petroleum and other imported fuels from world markets. We each depend heavily on fossil fuels—especially coal—despite the health and environmental problems that they cause. With regard to our use of commercial nuclear power we share a concern for costs, reactor safety, nonproliferation, and safe disposal of nuclear waste. We also face serious, although quite different, obstacles to the deployment of new energy technology: China's research and institutional infrastructure inhibits its progress, while lack of demand for new energy sources inhibits the United States. Both countries face the challenge of market reform, especially in the power sector, and both share an interest in technology cooperation—the United States in exports, the Chinese in imports.

In addition, China faces the multiple technical and economic challenges of becoming a mature energy economy, which, from the experience of other countries, requires access to commercial, high-quality energy resources, low energy intensity, increased electrification, and greater attention to environmental control.

The United States, on the other hand, is a mature economy confronting the special challenge of mitigating the spillover effects of its energy demand and technological choices on the global commons—notably in the demand for liquid fuels and the production of greenhouse gases—and on the choices of other countries.

In this chapter, our two countries' challenges and special needs are examined from their common roots, with differences noted in each broad area.

A. IMPORT DEPENDENCE AND ENERGY SECURITY

In the United States, concerns over energy security and import dependence are linked most closely to petroleum issues. The United States currently imports about half of its petroleum, and petroleum imports will rise to 65 percent by 2020. The Organization of Oil Producing Exporting Countries' share of those imports will rise to about 50 percent in 2020. By 2020 the Persian Gulf region will produce about half the world's oil, but imports from all regions are subject to concerns in varying degrees over transport and geopolitical implications.[20] China likely will import about 40 percent of its oil and about a quarter of its natural gas by 2020. Although concerns over increasing imports are a key driver of China's energy policy, energy security also means increased variety of fuel sources, especially diversification from China's heavy dependence on coal (see sections on "Coal" in Chapter 1 and the following section in this chapter).

China's petroleum industry is facing unprecedented difficulties in meeting the surging domestic demand. Aging onshore oil fields, below-expectation offshore production, and slow development of new oil fields in the remote western region all contributed to only modest growth of crude oil production over the past 15 years. China has been a net oil importer since 1993, and net imports were about 1 EJ (over 20 million tons) in 1996, most of them oil products. China imported about half its crude oil from the Gulf region in 1997 and, as imports increase, so likely will the share from the Gulf region. China will need to make significant investments in the refining sector to accommodate this increasing share of high-sulfur crudes from the Gulf region.

Chinese experts estimate that domestic crude oil production could peak at around 9 EJ (200 million tons) in 2020. The gap between demand and domestic supply likely will be 6 EJ (130 million tons).

With natural gas expected to represent 10 percent of primary energy supply in 2020, China will use about 200 billion cubic meters (8 EJ), more than half of which likely will be imported. Liquefied petroleum gas (LPG) also is increasing; between 1994 and 1996, imports increased more than threefold, from less than 1 million tons in 1994 to almost 3.5 million tons in 1996.

It is clear why the Chinese government has elevated energy security initiatives to a higher level of priority. Maintaining a strong domestic oil industry is considered by the government to be a partial safeguard against the unpredictable international market; achieving this goal without enduring too great a burden on the economy is a great challenge.

These considerations add emphasis to the importance of alternative transportation options. Options should include mass transit systems, more efficient vehicles, and alternative-fuel and low-emission vehicles. Infrastructure planning should be also considered in this context.

---

[20] Oil represents about 40 percent of total world energy supply and about 90 percent of energy traded between countries.

B. DEPENDENCE ON FOSSIL FUELS—ESPECIALLY COAL—
AND THE ASSOCIATED ECONOMIC, HEALTH,
AND ENVIRONMENTAL IMPACTS

The United States is the world's largest emitter of greenhouse gases, and 83 percent of all emissions come from energy production and use. By 2020 the United States will account for about 22 percent of total world emissions. Emissions from the U.S. energy sector are expected to rise by an average of 1.3 percent per year through 2020, from 1,480 million metric tons (Mt) in 1997 to 1790 Mt in 2020 (though relative share of world emissions remains constant). Fossil fuels, particularly petroleum (mostly used in the transport sector) and coal (used in power generation) are the major sources.

Under the Kyoto Protocol, the United States would be committed to reduce emissions of six greenhouse gases by 7 percent from 1990 levels[21] over the period 2008 to 2012. For the United States to actually meet the Protocol goals would require a radical change in energy policies or a major tradable emissions regime. The committee had not examined a detailed scenario for achievement of the goals. Possible emission reductions options include an emissions trading scheme and activities implemented jointly, perhaps through the Clean Development Mechanism (CDM) under which emissions credits would be received for projects undertaken in non-Annex I countries.[22]

A great deal of work has been done in the United States to quantify the external costs associated with energy production and use,[23] though there has been little success in incorporating these costs in the price of energy (see "Barriers to Deployment of Advanced Technologies and Practices" later in this chapter). Major energy-related problems include water quality, waste disposal, coal mine disturbance, radioactive emissions and waste, and air emissions, including health effects of particulates and acid rain impact on agriculture and buildings. Because air pollution control is of the greatest significance to the Chinese situation, a brief overview of the U.S. experience is provided for context.

Over the past 25 years the United States has succeeded in making significant reductions in air pollution through a three-part system of numerical standards, implementation agreements, and vigorous enforcement. Major components of the U.S. air pollution control system include limitations on SO_2, particulates, tropospheric ozone, lead (from gasoline), CO, and NO_x. Where elements of the

---

[21] For the three synthetic greenhouse gases, 1995 levels may be used.

[22] There have been several strategies developed to achieve emissions reductions in the United States beginning with the 1993 *Climate Change Action Plan*. Other efforts include *Technology Options to Reduce U.S. Greenhouse Gas Emissions* and *Scenarios of U.S. Carbon Reductions: Potential Impacts of Energy-Efficient and Low-Carbon Technologies by 2010 and Beyond.*

[23] See the report series prepared by Resources for the Future and Oak Ridge National Laboratory on estimating externalities of fuel cycles (ORNL/RFF, 1992-96).

U.S. program restricting these pollutants could be of benefit to China they are included.

U.S. SO_2 emissions have dropped by 8 percent since 1986, with corresponding benefits to health and reduction in acid rain. Emissions trading is a key element of this control system and one that might be of significant interest to China. U.S. particulate matter emissions dropped from about 13,000 tons in 1970 to about 4,000 tons in 1995 because of tighter regulations and technology requirements, with corresponding benefits to human health and regional visibility, another key concern to China. The health impacts from lead in gasoline are very well understood and lead as an energy-related source of environmental contamination has almost disappeared. Carbon monoxide, another pollutant coming from the transport sector, has been reduced in the United States through the introduction of catalytic converters. Nitrogen oxides are addressed through emission limit standards for fossil fuel plants as well as catalytic converters on cars. Carbon dioxide is not regulated or taxed and will need to be dealt with in the context of global emissions reduction targets.

China currently accounts for about 13 percent of total world carbon emissions, second only to the United States, and is widely projected to surpass the United States over the next several decades. The industrial sector in China—which accounts for almost half of total gross domestic product (GDP)—produces about three-quarters of China's SO_2, flue dust, and wastewater and about 87 percent of solid wastes.[24] In 1998 China expended about 1 percent of GDP on pollution prevention (China State Environmental Protection Agency, 1999). To date, China has achieved great success in decreasing water pollutants, so that total emissions actually have declined despite rapid industrial growth, but this is not the case with air pollution. To minimize emissions to the extent possible, given the projected growth rate of the Chinese economy, China will have to dramatically reduce air pollution intensity.[25] The consequences of China's serious air pollution are already apparent in human health impacts and decreased economic productivity.[26] In 1997, China began the process of phasing out leaded gasoline by 2000, starting in Beijing.

China addresses its current dependence on fossil fuels—especially coal—in three distinct manners: (1) decreasing coal use relative to total energy supply and increasing diversity and quality of primary energy supply, (2) increasing efficiency of energy production and use, and (3) moving to lower emissions limits on fossil fuel burning.

---

[24] A detailed examination of the impact of industrial pollution in China was prepared by the World Bank (1997b).

[25] Garbaccio et al. (1998) examine the effects of carbon taxes to reduce greenhouse gas emissions.

[26] For a detailed discussion of air pollution abatement strategies see World Bank (1994). For a broader analysis of the economic impacts of pollution in China see Smil and Mao (1998).

Perhaps the most significant of these is decreased relative dependence on coal resources, as the implications of coal use are far-reaching and cover economic, health, and environmental performance. The move to higher-quality and more widely available commercial energy requires the development and deployment of advanced technologies. The second initiative, increased efficiency of energy production and use, involves current and future technologies to decrease energy inputs while maintaining economic growth. The third, lower emissions limits, has immediate benefits and provides incentives for the other two.

Health, economic, and environmental considerations are some of the factors entering into the strong move to diversify fuel sources and decrease dependence on coal. In the time frame of this study, however, coal will remain the dominant fuel source, and even with major effort the reduction of dependence on coal will be only 75 percent to about 68 percent of total primary energy consumed.

Given the magnitude of this medium-term burning of coal it is important for all countries involved to begin exploring approaches to sequestering CO_2, and the committee strongly encourages these collaborative activities. Carbon sequestration is a real possibility, especially in a scenario in which a short term reduction in CO_2 is required. The U.S. Department of Energy is currently expanding its programs in this area and there are interfaces with coalbed methane, enhanced oil recovery, and geologic disposal of CO_2.

Hydroelectric is an example of the effort to diversify fuel sources in China. In recent years it has contributed greatly to the increased electricity capacity in China and is projected to account for almost 70 GW of capacity by 2000 and 160 GW by 2020.

China has already made impressive gains in the second activity of increasing energy efficiency, reducing energy intensity by 50 percent over the period from 1980 to 1995. Tremendous gains have been made, but the opportunities for improvement are still vast. The recent passage of the Energy Conservation Law provides the opportunity for China to promote the introduction of energy-efficient technologies in China in numerous ways.

Natural gas has replaced coal in some of the most polluting applications—especially direct coal burning in the residential sector for cooking[27]—and is well positioned to make a strategic contribution to China's energy sector in the 2020 time frame. As in the case of oil, the market conditions do not yet fully support economic development and exploitation of natural gas, nor are the institutional mechanisms in place.

China is also undertaking improvements in energy infrastructure to support higher-quality energy systems. Notable efforts include recent natural gas pipeline projects, and plans to integrate its regional electric power grids (see "Energy Infrastructure" later in this chapter).

---

[27] Direct use of coal and firewood for heating is still a common practice; about 90 percent of China's population rely on solid fuels.

China has recognized the need to revitalize its pollution levy system to make it more responsive to changing market conditions, to include more pollutants, to raise its penalty rates to provide necessary incentives, and to allow some flexibility in how provinces handle pollution controls.[28] When fully implemented, these changes will bolster China's ongoing efforts to introduce coal washing and other clean coal technologies.

C. NUCLEAR POWER CHALLENGES AND OPPORTUNITIES

Opportunities for nuclear power are distributed broadly around the world, and nuclear power can help to address concerns over sustainable energy resources. This is very significant for the United States and China, whose dependence on coal is a shared concern and in a world now considering limiting the release of carbon dioxide. The challenges of nuclear power are widely recognized to be four: cost (particularly large initial capital costs), operational safety, the safe disposal of nuclear waste, and the prevention of the proliferation of nuclear weapons.

The nuclear programs in China and the United States face specific challenges and opportunities in the years just ahead, but the contrasts between them are considerable. The U.S. nuclear program of about 100-GW generating capacity is the largest, most mature in the world, but one which suffers from public neglect and lack of governmental support despite a continual improvement in plant performance and safety in recent years. China, in contrast, has only recently launched its nuclear program, which now includes a 300-MW Chinese pressurized water reactor (PWR) and two French 900-MW units. China views nuclear power as an important alternative to coal among its electric power generation resources. A closer examination of the two programs will help to demonstrate the magnitude of our shared interests and the importance of collaboration.

The U.S. problems are closely associated with the Three Mile Island accident, which led to the decline of public support for nuclear power, with environmental groups being particularly outspoken. Although the plant was a total loss, there was essentially no offsite damage or injury. The later Chernobyl accident, in contrast, occurred at a plant without a containment system and led to widespread radioactive fallout, loss of life, and damage and contamination throughout the plant, the nearby countryside, and to the surrounding region. This event rocked Europe and the Soviet Union and further compromised the future of nuclear power in the United States and many other countries.

Asia was less affected by these events, in part because nuclear power did not play a prominent role in countries other than Japan, but the lesson for Asian nations now turning to nuclear power should be clear: public support is fragile

---

[28] For a detailed discussion, see Chinese Research Academy of Environmental Sciences (1997).

and cannot be taken for granted. Safety is an imperative in nuclear operation and, as the United States has learned, an accident anywhere affects all operators in our open and networked societies. Through the Institute of Nuclear Power Operations and the World Association of Nuclear Operators, these and other lessons learned have been shared among U.S. and other operators. Although safety records have been good, plants have experienced problems affecting reliability and plant economics that need to be shared with new operators, including China. Generic problems, such as steam generators in PWRs have been a major problem.[29] Much has been learned not only by the owners and operators, but also by regulators, environmental and financial communities, and governments at all levels. China can benefit from these lessons as it launches its nuclear era.

China recognizes the growing greenhouse gas concerns around the world and is paying increased attention to both CO_2 and methane emissions. Nuclear power presents many advantages to China as a complement to coal, which, as a relatively remote resource in China with respect to population centers, has created serious congestion problems for rail, road, and water transportation systems. In particular, nuclear power is an attractive option for meeting the energy needs of the dense population centers in eastern and southern coastal regions. Nuclear power as an alternative to coal use can ease Chinese environmental problems as well as global concerns.

However, as noted earlier, nuclear power is not only technically sophisticated, but is also capital intensive. The latter factor remains a challenge that has led China to seek investment as well as to assess options beyond its own PWR design.[30] In addition to French PWRs, China has entered arrangements with Canada for two 700-MW CANDU reactors and with Russia for two 1,000-MW PWR units.

China intends to continue development of its own PWR design now targeted for 1,000 MW as its highest-priority effort in nuclear power. China probably could benefit from both the U.S. advanced PWR design and the French-German advanced PWR at 1.5 GW. China's approach can be informed by the successful French program that standardized and periodically upgraded designs but all units were identical in each generation. Customization of units for the U.S. market by global vendors contributed to problems, some operational, but also economic. China's plan calls for major projects using the Chinese design at the 300-, 600-, and 1000-MW levels for development early in 2000 with 20-30 GWe operating by 2010 and 40-50 GWe by 2020.

The U.S. outlook is complicated by the ongoing restructuring in the electric industry. As the early plants approach the end of their 40-year licenses, decisions

---

[29] Some advanced nuclear designs will not suffer from this problem, as they have eliminated steam generators entirely.

[30] Other advanced designs, including second generation helium-cooled plants offer significant improvements in efficiency, safety, and resistance to proliferation.

must be made with respect to license extension. Units must pass an inspection, and many likely will require additional investment to qualify. Decisions likely will be made on a plant-by-plant basis, with many single-plant owners likely to sell to operators who hope to lower average unit costs by expanding their interests in nuclear power. Plants likely will face retirement when plant performance or high projected cost assessments indicate that they cannot compete economically. U.S. natural gas prices projected for the next 20 years will make natural gas plants very attractive alternatives to new nuclear orders. However, a significant carbon tax or CO_2 emission target would give an economic advantage to nuclear power, particularly in comparison to coal, but also, to a lesser extent, to natural gas.

The challenge of preventing proliferation is one faced by the United States, China, and the world community in general. Thus, common priorities are for research and international institutional cooperation on protecting, controlling, and disposition of fissile material. The challenge of safe disposal of nuclear waste is also one of common concern, about which China, the United States, and other countries can learn much through cooperative efforts.

The commercial nuclear power industry in China and the United States would benefit greatly from mutual cooperation: A rejuvenated market for nuclear technologies will lead to faster development of better designs and increased technological capability. In the short term the United States stands to increase exports of equipment, and China would benefit from increased power capacity and transfer of technology. An extensive nuclear program implies the need for measures—both technical and institutional—to ensure that the threat to security is not increased. As noted in Chapter 1, a framework agreement that would allow this type of cooperation is already in place between both governments.

D.   RENEWABLE ENERGY SYSTEMS

The following is a brief overview of the status and trends in the renewable energy area in both China and the United States. Because the market, institutions, and technology for hydropower development are well understood and mature in both countries, this discussion focuses on biomass, solar, and wind, renewables that have promising futures, but are facing special barriers (such as implications for land use, intermittent availability, or need for storage) in their ascendance to become efficient and economic energy sources for both countries. Barriers specific to China are presented in Box 2-1. In the time frame of this study renewable energy technologies are important in a strategic role, often in conjunction with fossil fuels, conventional hydropower, and nuclear energy, or in remote applications if cost and energy storage problems can be successfully addressed. In their present state of development they are not a solution to large-scale energy supply.

China and the United States have among the world's largest endowments of

BOX 2-1 Challenges to China's
Renewable Energy Development [a]

China's renewable energy program, although sufficiently coordinated at the central government level, is undertaken by relatively isolated small research and design institutes or local government departments, and has been met with a host of challenges in its effort to grow:

- Institutional fragmentation and uncoordinated project efforts hinder the development of market-oriented commercial companies, and hamper effective networking with commercial/financial establishments and private and foreign investors.
- Insufficient market regulation and the lack of industry standards suppress demand because of widespread product quality and service problems.
- Weak linkages with the financial community and underdeveloped government financial policies/regulations result in a dearth of credit and venture capital for even worthy projects and commercial applications.
- Insufficient commercial experience prevents the renewable energy industry from delivering its products and services efficiently.
- Awareness of potential and opportunities for commercial applications is lacking among decision makers and interested parties outside the renewable energy community.
- Information exchange is lacking among practitioners and access to advanced technologies is insufficient.
- Insufficient assessment of resources affects investment decisions and project finance.

China also faces major technical problems:

- Applications are small and scattered, and suppliers of equipment are saddled with outdated manufacturing technologies and are incapable of achieving economies of scale.
- Funding for R&D is inadequate, partly because of low demand, and relative isolation from the international community hinders technological advancement.
- Inexperience in project design and implementation and the lack of market orientation increase construction costs and cause operational problems.
- A clear strategy is needed on how best to combine China's comparative advantages and international state-of-the-art technologies to further long-term development goals and how best to reduce costs.

---

[a] Adapted from a Chinese Government and World Bank study on China's renewable energy development strategy (World Bank, 1998).

biomass, solar, and wind resources. Large-scale deployment of advanced applications of these renewables may be an important long-term energy strategy for both countries because they are alternatives to fossil fuels and produce zero net carbon emissions. However, for these renewables—especially solar and wind—to become significant sources of energy, each country will have to overcome its own set of challenges. This effort could benefit greatly from close cooperation.

Reducing costs and nurturing the market for renewable energy technologies are significant tasks for both countries, but China faces particular difficulties. It lags behind the United States in technology development and manufacturing capability and has the double duties of developing the market infrastructure for renewables while making the transition to a market economy. Sustaining and expanding the market development of each renewable energy technology also has its own special challenges.

Biomass

Biomass electricity generation is the largest source of non-hydropower renewable energy in the United States, with a capacity of about 10 GW. Forest products and agricultural residues and waste long have been used for cogeneration of electricity and heat on the order of 7 GW. Several U.S. electric utilities also have demonstrated in recent years that these products can be used with acceptable technical and economic performance in specialized conditions—in larger size plants with greater operating efficiencies, when both steam and heat are needed and when fuel costs are attractive.

Biomass co-firing is the burning of biomass fuels with coal in existing plants without the need for new boilers or gasification systems. Significant benefits of co-firing include reduced plant emissions and disposal of a waste product.

Biomass gasification plants in the United States currently being developed and demonstrated might offer the greatest benefits: higher thermal efficiency, scalable applications from 5 to 100 MW and increased fuel flexibility (EPRI/DOE, 1997). Municipal solid waste and landfill gas plants in particular have promising futures as efficiency levels rise with combined-cycle technologies, if costs can be decreased for both fuel and the production facilities.

In the longer term, biomass gasification fuel cell systems might be an attractive application in the United States and abroad, especially in smaller applications as part of a distributed energy strategy.

The importance of efficient stoves notwithstanding, the primitive and inconvenient use of solid biomass in rural China has been declining steadily as modern fuels, including coal and LPG, have made great inroads in the past 10 years. Disposing of surplus crop stalks by burning them in the fields has become a serious problem in much of the northern rural areas, causing acute local air pollution, and serious visibility reduction. Biomass gasification is in an early stage of development in China. Several research institutes design, manufacture, and market gasifiers, but these are scattered efforts that fall far short of building a commercial infrastructure for a potentially huge market in efficient modern use of biomass. China has perfected anaerobic fermentation technologies in its large-scale rural household biogas program and has used them to treat a limited amount of industrial wastewaters. Increasing the use of anaerobic digesters in industries such as distilleries, sugar, and pulp and paper would generate great environmental benefits while producing energy. Bagasse cogeneration has a potential to add 700 to 900 MW of cost-effective generating capacity to the grids in China's sugar-producing provinces. However, technical as well as financial difficulties have restricted bagasse cogeneration to in-mill use only.

Solar

Solar photovoltaic (PV) electricity generation is used in niche, off-grid applications where its strengths—versatility, reliability, and absence of harmful emissions—outweigh its significant economic cost.[31] The need for storage systems or backup options to address intermittent availability is also a significant challenge. The transition of PV technology to widespread more economically competitive applications, however, is not entirely clear, and grid-competitive PV electricity in the United States may lie outside of the 2020 time frame for this study. A new trend in PV deployment—building-integrated, in which PV equipment is included in the actual building materials—may be the key to decreasing installation costs. This approach has received significant attention in the U.S. government and in other governments elsewhere.

Solar thermal energy technologies were first deployed in the United States in the 1980s under a program of state and federal incentives to foster development of emerging renewable energy technologies. Over the past decade, considerable experience has been gained with these technologies, and costs have decreased significantly. Solar thermal in combination with natural gas may be an attractive option in distributed applications where there is a rapidly growing need for electric power.

Quality of products and customer services are a common problem of solar applications in China and undermine the efforts to increase market demand. Market demand also is affected by a lack of mechanisms to provide consumer credit in a high capital-cost application. The production scale of PV cells and modules in China is small, contributing to higher prices. Production was less than 2 MW in 1996, compared with a total shipments of 35 MW in the United States for the same year. China's solar thermal collector industry is fragmented, with small and outdated production lines and is insufficient to meet demand in

---

[31] Relative to conventional grid electricity, solar PV electricity is about 5 to 10 times more expensive (EPRI/DOE, 1997).

quantity and quality, especially for large-scale commercial and industrial applications.

Wind

Wind power systems in the United States have benefited greatly from government incentives over the past 20 years, and costs have continued their downward trend. In the past few years, wind technologies have neared the point at which they can provide competitive peak power. The U.S. experience has been mainly with large wind farms connected to a transmission grid through a dedicated substation, though interest is increasing in distributed facilities in which units are connected to a utility distribution system, a common model in European wind applications.

China's efforts to tap its large wind power potential are limited by its lack of widespread technical and financial know-how. Domestic manufacturers are capable of producing 100-W to 5-kW mini turbines and have begun to produce 200-kW turbines, but larger and more efficient turbines must be imported, and high costs have limited their use in demonstration projects. The wind power industry also lacks experience in design, construction, and operation of large wind farms. The government has yet to promulgate clear and conducive policies/regulations to make large wind power investment financially attractive to utilities or independent power producers. These are areas in which the United States has had substantial experience and could provide crucial assistance.

E. ENERGY INFRASTRUCTURE

The energy infrastructure that links energy resources to customer energy services involves extraction, processing, conversion, waste disposal, and transportation through pipelines, railways, waterways, roads, ports, and electrical grids. The U.S infrastructure is well developed for a fossil-based energy economy with rails, pipelines, and roads providing the major transport for coal, oil, gas, and refined fuels, while electricity moves to market over high-voltage grids. China's developing infrastructure has components similar to those of the United States but is less developed at this point and its components have different relative importance.

These modes of energy transport probably will not continue to offer optimal efficacy in the future. Already both the United States and China—the largest transporters of coal by rail—are experiencing congestion on the railroads. The inability of the U.S. government to agree on how to fulfill its obligation to accept waste from U.S. nuclear plants is one of the factors compromising the viability of nuclear power in the United States. The electric grid in the United States generally has provided reliable service, but not without controversy in regard to environmental and aesthetic concerns and occasional disruptions ranging from power quality issues to outages. Recent trends in deregulation of power supply activities are creating new challenges for transmission and distribution system control. The technology behind the moving of energy must be improved to lessen environmental concerns and improve the efficiency and operability of these systems, particularly in electric power.

Although it seems reasonable to assume that superconductivity, flexible alternating current transmission (FACTs), better insulation and undergrounding capabilities can help to improve electrical grid performance and acceptance, some new technologies such as distributed generation, including renewable sources like solar and wind, can bypass all or much of conventional transmission and distribution (T&D) systems. Distributed generation involves small generators— 25-kW to several-megawatt gas turbines, internal combustion engines, or fuel cells—that offer onsite generation potential. These systems do not rely on the transmission grids that have been the backbone for getting today's central power station coal and nuclear electricity to the distribution networks. Rather, they are able to use natural gas or liquid fuels and generate power close enough to customers that waste heat utilization becomes more attractive. Such systems will offer opportunities to both China and the United States—particularly in rural areas where service is less developed.

Similarly, renewable resources such as solar, wind, biomass, and small-scale hydro often can be sited at or near loads so as to bypass the high-voltage grid, though it is advantageous to provide larger energy production sources with grid access to ensure full utilization of power production during optimum conditions. Solar photovoltaic and biomass both offer localized energy sources and can be combined with either storage (which needs more development) or two-way grid connections. As increasing attention is given to distributed generation, and as long as gas prices remain attractive, the energy transport investments likely will favor pipes over wires. Renewable-fueled distributed generation also could lessen dependence on electrical grids if energy storage technology improves. The increased use of distributed energy sources in nodes of concentrated energy use also has possible implications for the reliability of the balance of the interconnected system.

However, for countries or regions heavily dependent on coal, mine-mouth power has certain advantages: It permits generation in remote areas where it may be possible to localize waste disposal and sequester carbon, and it reduces rail congestion and urban pollution. Coal-by-wire would require a cooling water source and decreases the number of cogeneration possibilities.

These points deserve attention as both the United States and China upgrade their infrastructures. Mine-mouth plants could offer significant environmental benefits, and today's high-voltage grids offer improved delivery capability. It will become increasingly important that future system planning take these factors into account to ensure that existing plants and future sites will offer attractive investment opportunities for generators, including opportunities to market waste heat and minimize environmental costs. Similarly, current design nuclear plants require remote sites with adequate cooling, grid-access, and adequate spent fuel storage space.

In the United States, new gas combined-cycle electricity generators are being built by independent power producers where grid access maximizes access to markets. Similar considerations will apply to investors in China, and it will be important to discuss both pipeline and grid configurations with potential power plant investors, as new investments are planned.

China built its first gas transportation main line in the 1960s, and by 1996 had a total network of about 3,700 km (of pipe diameter over 426 mm). Transportation of commercial natural gas reached 10.4 billion m^3 per year in 1996. China's natural gas pipeline transportation industry is still in its early stages: The geographic distribution is uneven, and a national system is incomplete. Low utilization of pipelines is due in part to the following situations: traditionally oil exploration has taken precedence over gas exploration, with corresponding development of infrastructure; historically, the trend has been to connect a single gas source to a single user, rather than connecting to a network; and there has been a lack of gas storage facilities to adjust peak demand. In order to vigorously develop natural gas, it is necessary first to invest in a modern pipeline system, and plans are in place to do so by 2010.

About 40 percent of China's national rail capacity is devoted to coal transportation. The system is heavily stressed and upgrades are extremely expensive.

The status of China's electrical grids has already been discussed. The upgrading of this system is an urgent priority.

F. BARRIERS TO DEPLOYMENT OF ADVANCED TECHNOLOGIES AND PRACTICES

The deployment of new energy technologies on a large scale is a major challenge to governments in both China and the United States, though the specific situations in the two countries are somewhat different. Rapid economic growth has created an urgent priority to meet rising energy demand in China, which, coupled with the movement toward market reforms and decentralized political as well as economic spheres, makes for a complex task for government entities at several levels. The downsizing and decentralization in the Chinese government is occurring on an unprecedented scale. While these are complex tasks, restructuring is taking place.

Deregulation of the Electric Power Industry

The United States has a much more developed energy infrastructure than China, though the issues of privatization and deregulation are resulting in some sweeping changes in how this country meets its energy needs. Many have been concerned that deregulation will cut back or eliminate efforts such as utility-driven demand-side management, integrated resource planning, and load management programs and thus result in an increase in energy demand. Others suggest deregulation simply will shift the incentive to save energy to the end user, as evidenced in the growing energy services company (ESCO) industry in the United States and in Europe. Implementation of international global climate initiatives, however, could provide strong incentive to further develop competitive energy-efficient and renewable energy systems.

Deregulation of the electric power sector in the United States has the potential to negatively affect the development and deployment of competitive renewable energy systems and energy efficiency. Efficiency goals have been driven largely by utilities, and increased application of renewable energy systems traditionally have been promoted by regulators. In a deregulated system where price of electricity is the primary factor, renewable energy systems at their current level of technology can compete only in niche applications, usually in remote areas off-grid or in highly specialized applications. Demand side energy efficiency may not be a priority to a power marketer if the result is to limit sales of its products and services.

A further concern is that private research and development (R&D) funding for commercially competitive renewable systems also could decline if current price of power were the only driver. Utility-driven R&D programs also would tend to focus on near-term goals, rather than a long-term strategic objective. Thus, the challenge will be to provide the proper incentives to develop economically viable renewable energy systems without re-regulating the industry. The role of government-funded R&D remains important for precompetitive technologies in the deregulated electricity markets.

Within the context of deregulation, however, some attempts have been made to shape the industry in the United States. The Renewable Portfolio Standard (RPS)—mandating that a certain percentage of a utility's electricity be generated or purchased from non-hydro renewable sources—is being implemented in some states and may help to secure modest contributions from renewables to grid-based systems. Such a mandate increases interest and investment in renewable energy systems. The potential positive environmental impacts in the entire United States could be significant: A contribution of 5.5 percent of electric capacity from non-hydro[32] renewable energy would account for a reduction of over 20 million metric tons of greenhouse gases in 2010 (DOE/PO-0059, 1999). The creation of this standard is controversial and likely will lead to considerable political debate within the U.S. government.

---

[32] U.S. electricity generation from conventional hydropower decreases by 0.3 percent per year through 2020 and thus is not included in this account of CO_2 reductions from renewable sources.

Chinese electric power reform also is taking place at a rapid pace. The Ministry of Electric Power has been replaced by the State Power Corporation, a nongovernmental body that owns half China's electric power generation facilities and all of the T&D network. The next step is the separation of generation from T&D and the creation of competitive markets. Three test areas—the independent Sandong power grid whose installed capacity in 1998 was over 17 GW, Zhejiang power grind and the Shanghai power grid within the East China Power Network whose installed capacity in 1998 was over 46 GW, and the North-East China Power Network with a capacity of over 37 GW in 1998—have been selected to demonstrate an independent power market; power rates in these areas are expected to decrease and service quality improve. If these test areas proceed as planned, a wholesale power market could be created in China as early as 2010, the year in which China plans to complete the interconnection of its six regional and five independent power grids. In the three test areas for deregulation there is no provision to promote cleaner energy sources, nor are there penalties for heavily polluting electricity suppliers.

Investments and Market Reforms Needed To Promote Advanced Technology Deployment

Investments in energy technology improvement may be categorized into three types: large investments in new facilities and processes; large investments in retrofitting existing facilities and processes; and small investments in new or existing facilities. The first type may include investments in new production processes, new power plants, and new buildings and vehicles. Embodiment of advanced and energy-efficient technologies and designs in new construction is very important in China's energy conservation efforts, considering that most of the capital stock for 2020 is yet to be accumulated.

The second type involves investment in modernization of existing facilities, such as upgrading production processes, rehabilitating old power plants and T&D systems, and improving coal preparation. Because there is enormous potential for cost-effective improvements in energy efficiency—especially in the industrial sector in China—such investments are crucial in the short to medium term and would be recurring activities in the long run as long as they have positive economic effects.

Energy efficiency per se often is not the main purpose of the above two types of investments, which are relatively large and are usually intended to expand production, increase productivity, improve product quality, or decrease emissions. Special policy incentives are needed to create more favorable market conditions for private investment in the deployment of energy-efficient technologies.

The third type includes a variety of relatively small investments, including personal investments—primarily for saving energy—such as measures to improve small boiler combustion efficiency and recover waste heat, as well as installing efficient lighting equipment, or spending additional money to purchase more energy-efficient refrigerators. The energy savings of individual investments tend to be small, but the aggregate savings can be substantial and the payback period relatively short. Many of these investments have quite high returns, but constraints to implementation are sometimes quite strong—mostly high up-front capital costs and transaction costs.

Market-oriented reforms and rising income have greatly changed the incentives and constraints to investments in advanced energy technologies for enterprises and individuals alike. The Chinese government recognizes the need to improve the effectiveness of policies and strongly supports the development of market-based initiatives; energy prices now are largely market-driven, and ESCOs are being introduced with assistance from the Global Environment Facility and the World Bank. Regulations have been strengthened to correct market failures. Air pollution controls have been tightened, and in 1997 China passed a comprehensive Energy Conservation Law. Continuously adjusting and updating energy policy approaches within changing overall economic systems is important to maintaining the momentum of energy improvements in China's transitional economy. The challenges to implementing cost-effective energy programs, and policies should not be understated.

Barriers to Investment

Despite the progress made, there are still major barriers to investments in advanced energy technologies, many pertaining to institutional and regulatory reforms necessary to create the proper framework for energy projects. Specific examples of these barriers for China can be found in Box 2-2: They are presented in the context of energy efficiency investments, but many have broader significance.

Externalities and the Failure of Energy Prices To Reflect the Full Cost of Energy Use

In a market-based economy, one function of government is to protect the public from the actions of parties who are motivated only by the private costs that are reflected in market prices—at least when the divergence between private and societal costs is large and the matter is of major significance. These conditions can exist with regard to energy production, conversion, and consumption; absent government action, these processes may be "underpriced" when their full costs are taken into account. The difficulty is in finding acceptable methods to accomplish this objective that are practical and that do not cost more than the conditions that they are designed to remedy.

First, the benefits of energy use are manifest and the unpaid costs sometimes are hidden, in the future, and/or widely dispersed. These make the political diffi-

BOX 2-2   Barriers to Investment in Energy Efficiency in China

- **Unfinished economic reforms**, especially those affecting productivity and investment decisions, such as state-owned enterprise reform and financial sector reform.
- **Underdeveloped policy framework** for market-based energy efficiency incentives. Many of the administrative measures previously used to promote energy conservation are becoming inapplicable in a changed economic environment.
- **Insufficient institutional capacity** to support market-based energy efficiency initiatives. China's existing energy conservation system, although extensive, was originally set up to help the state-owned industries, and has not changed adequately to provide the kind of push and support that enterprises and individuals may need in a market-oriented economy.
- **Underdeveloped** industrial/building **standards** and related testing, monitoring and enforcement capacity.
- **Lack of effective policies/mechanisms** to improve the technologies of China's many small industrial enterprises, which have been the driving force behind China's industrial growth.
- **Lack of information about the financial aspects** of energy-saving investments and about opportunities and best practices.
- **Inadequate technology dissemination** due to the lack of effective mechanism and insufficient efforts.
- **Small energy (and cost) savings and high discount rates** often override the long-term benefits of individual energy efficiency investment, inhibiting the implementation of many cost-effective measures.
- **Real and perceived high transaction costs and risks**. Reliability and consistent performance is highly valued in China, discouraging enterprises or people from trying unfamiliar new products and new technologies, especially those without reputable names.
- **Difficulties in arranging financing** because banks are unfamiliar with the financial aspects of energy efficiency investments.
- **Insufficient internalization of the health and environmental costs of energy consumption**, especially those associated with coal production and use.

culty of gaining acceptance of such actions very serious. Even when such externalities are clearly manifest, e.g., urban air quality in the Los Angeles region a decade ago, they are difficult to address.

Second, instruments to intervene to address externalities are inherently difficult to precisely implement with regard to cost and to implement in ways that are transparently fair. Indeed, they will always fail the test if absolute equity and efficiency are the standards—the best that can be achieved is to get the system approximately right, with the minimum disruption to market forces. Therefore, not only must difficult actions be taken, not only will their benefits be somewhat veiled, but they also will be difficult to defend in particulars as meeting tests of fairness and efficiency.

Third, such interventions themselves are costly and a drain on the economy. These costs include the resources absorbed in creating, administering, complying with, and enforcing the regulations and must be considered along with the costs of the externalities themselves. They also include the "cost" in political will required to decide upon, gain acceptance of, and then carry out what always seem to be highly unpopular measures such as fuel taxes or increases in electric utility rates. In addition, required measures inevitably will lead to unpopular economic dislocations, for example, from reduced employment in some geographic areas and industrial sectors. The impact of these dislocations can be reduced if taxes are introduced gradually.

These interventions are costly because of mistakes and failure to perform on the part of the authorities imposing the interventions. The "market failures" that occasion intervention are mirrored by the "government failures" that make the interventions themselves less than optimal.

Finally, social objectives are constantly changing, as new information becomes available and as public priorities change, for example CO_2 emissions were not on the policy agenda 20 years ago.

The conclusion is that rational energy pricing for both China and the United States is both difficult to achieve and important to accomplish; perfection is not attainable. Governments have an array of methods to minimize environmental and safety hazards, including laws, regulations, permitting, and taxes. In the United States this range of actions has succeeded in internalizing many of the costs associated with electricity production. The United States has done particularly well in reducing SO_2 and particulates and has made progress in reducing NO_x.

Addressing externalities associated with energy production and use, however, is a continuing challenge. The important matter is to reduce gross distortions such as those that lead to large costs from local and large-scale environmental damage. Further, it is critical to avoid well-intentioned but destructive policies that subsidize energy use to achieve other goals such as improved personal income distribution, industrial development, or export promotion. Beyond that, the goal should be policy choices that are robustly directionally correct, within the context of a free market, even though they cannot be fine-tuned to cover all of the "unpaid costs" of energy production, conversion, and consumption.

China has been experimenting with and developing approaches for internalizing costs. This committee applauds these efforts and supports continued progress.

G. GLOBAL RESTRUCTURING OF THE ENERGY INDUSTRY

The role of government in the energy sector has been undergoing rapid change in recent years throughout the world. Many centrally planned economies are turning to market systems to meet the needs of underserved and growing populations. Developed countries with well-established markets and private ownership are finding that deregulation of gas and electricity production and sales offers the promise of significant economic benefits. Only the T&D systems remain regulated.

As these changes progress governments are rewriting the rules of engagement to ensure open competition and transparent markets, to attract investment, and to provide reliable, affordable services to customers in all markets. These changes are occurring in both China and the United States at varying paces.

In recent years some Chinese state-owned enterprises have been restructured to perform better under reformed market conditions. Private investment is increasing rapidly in energy supply, much from international ventures. Given the size of China's unserved population, a continuing government presence will be necessary to ensure that commercial energy services reach outlying areas, and China's policy is to increase the availability and quality of energy services to all of its population as rapidly as possible.

The World Bank and other international financial institutions will play an important role in developing essential infrastructure to entice service providers and fuel and equipment suppliers to develop projects in new markets.

In the United States, deregulation eventually will eliminate monopoly providers and create competitive markets for both generation and power market services. Healthy cash flows and repositioned capital is being used by U.S. utilities to build critical-mass investments around the world in both generation and market services, two quite different businesses. These opening markets are attracting oil and gas companies as major players as well. Recently a major oil company and an engineering/architectural firm announced a partnership that likely will become a major global generation provider. Others already exist and many more will follow as the synergies between refining and generation are pursued. At this stage of deregulation the consequences seem possible, but these developments are subject to a variety of changeable factors.

Utilities, like oil companies, have split their businesses into upstream and downstream components as technology services become increasingly specialized and important.[33] Advances in both seismic sophistication and directional drilling technology have revolutionized the upstream portion of the oil and gas business. Because of mergers and acquisitions, fewer global companies likely will dominate this part of the business, and they will continually compete to ensure that

---

[33] As noted earlier, China's oil industry has recently integrated upstream and downstream companies.

they have cutting-edge technology. Increased partnering among established and emerging players provides financing and risk sharing as they move offshore and into deeper water.

The refining and marketing sector is consolidating as it enters more competitive markets with tighter profit margins and more stringent environmental regulations. These changes could have an impact not only for the United States, but for China as well. Electricity providers, much like oil and gas companies, are seeking global opportunities. Mergers, acquisitions, strategic partnering, and alliances by which companies seek to position themselves to use technology resources and access markets are taking place at an accelerating pace around the world. China's opportunities to partner with global generating companies to provide optimal use of their natural resources may provide very valuable benefits to the developing Chinese utility system.

3

Conclusions and Recommendations

An appropriate strategy for meeting the energy-related challenges facing China and the United States would include several types of initiatives:

- Promotion of U.S. and Chinese investments in frontier technology important now for the United States and soon to be important for China. This would include advanced power generation, greenhouse gas controls, environmental technology, and transportation efficiency.
- Development of collaborative programs to accelerate the deployment of advanced technologies. There are significant opportunities in the oil and gas sectors, efficient and cleaner coal use, electrification, energy efficiency, and environmental controls. Priority should be given to technologies whose demonstration and deployment are feasible either in China or in the United States. Priorities should follow market conditions and energy sector needs, and programs should include both institutional and specific technical initiatives. The rapid development of the Chinese energy sector compared to the relative maturity of U.S infrastructure creates many opportunities for both countries.
- Ongoing collaboration between key scientific and engineering institutions in the two countries, notably the Academies of Sciences and Engineering, to help guide the choices required for implementation of these initiatives.

These types of initiatives are presented in this section broken down by sector, as discussed in Chapter 1. The findings and recommendations presented here are those of the Committee on Cooperation in the Energy Futures of China and the United States (CCEF or the committee) and are intended for institutions in both countries.

A. CROSS-CUTTING INITIATIVES

China and the United States are large and influential countries, and the sum of their energy production in the next century will represent a major portion of world total. Energy issues and especially the implications of increased energy demand and use are of great concern, not only to each country but to the global community. Successful cooperation between China and the United States will help the energy industry in the world to develop along a more sustainable path.

Many significant recommendations and initiatives for cooperation have been presented here for different fields, and both Chinese and U.S. experts have highlighted the technical areas in which opportunities might exist. There are many relationships and mechanisms for collaboration and the following institutions should be involved:

- government departments, ministries, and agencies;
- academic, scientific, nongovernmental, and trade association entities;
- private industry collaboration in the context of an expanded and deregulated commercial regime; and
- multilateral development banks.

Continued Cooperation Among the Academies

To promote cooperation between China and the United States, the Chinese Academy of Sciences, the Chinese Academy of Engineering, and the U.S. National Academies could help to sustain programs on the new and ambitious recommendations and initiatives suggested by the four Academies and others.

A1) The Committee on Cooperation in the Energy Futures of China and the United States (CCEF) recommends that a standing committee be established among the four Academies to identify opportunities for research, development, demonstration, and deployment of cleaner and more efficient energy technologies. The Academies are well suited to this task because they maintain strong contacts with government, industry and the international lending community and have the capacity to evaluate the technical merits of particular energy approaches and the framework necessary to implement them. A mix of government and private support would need to be found to support such a standing committee.

Such a standing committee should consider in particular the policy, regulatory, and incentive structure necessary to support clean and efficient energy production and use, one that clearly reflects external costs. Continued interaction on these issues is necessary because particular market conditions change rapidly as economies develop. Even if technologies well suited for Chinese markets are readily available, the institutional and regulatory environments may not be conducive to their use. To address barriers such as enforceable environmental requirements, confidence in protection of property rights, market pricing for energy, and advanced technology deployment, the Academies could provide expert input to policy makers on market and environmental reforms. The Academies could work with the World Bank, the United Nations, and others who are supporting reforms to ensure that advanced energy technology deployment is a part of economic reform programs. The Academies also could conduct seminars and expert analyses of these programs.

A1a) The CCEF recommends that the four Academies create a subcommittee to collaboratively assess the design and implementation of energy efficiency policies and programs. For China, this group would need to consider the design of strategies to implement the recently enacted Energy Conservation Law. For the United States this group could assess a range of opportunities to strengthen energy efficiency policies and programs. Such a continuing collaboration should include public and private-sector participants (including legislative and administrative branches; federal, state, and provincial officials; and academic and research institutions) with a strong knowledge of energy conservation policies and practices.

A1b) A subcommittee on the health impacts associated with energy production and use should be established—to include representatives from the Institute of Medicine and counterpart Chinese institutions—to provide quantitative input to decision makers in each government concerning health impacts associated with energy choices. This subject deserves early attention to address efforts to minimize the health impacts of energy use—particularly indoor air pollution in rural areas and the impact of the transport sector in urban areas—and could provide timely advice to agencies in both countries already working on this problem.

A1c) The Chinese and U.S. Academies of Sciences and Engineering could provide an "Academies-Industry Forum" to convene top-level energy industry representatives, researchers, and government agencies from the United States and China. They would discuss issues relating to policy and regulatory decisions and project guidelines, and their resolution to permit acceleration of investment and progress in one or more specific energy sectors and regions in China. The goal would be to experiment and to try to set an example of what can be done if clear and stable policies and guidelines are available. Specific local and regional circumstances (e.g., coal quality and specific development goals) would be taken into account.

Continued Cooperation Between Our Governments

It is recognized that before technologies are widely deployed throughout the world, several plants must be built to reduce the risks and prove their economic, technical and environmental performance and particularly to bring down costs through learning. This technology introduction and maturation period can last decades. The risks and costs of such commercial demonstrations may be more than most companies are willing to consider.

One method for accelerating the commercial introduction of the technologies is for governments to provide incentives to those willing to take the risks of new technology. Incentives can take the form of grants, low-interest loans, tax breaks, and other mechanisms. An insurance pool to cover new technology "make-good" should be also considered to facilitate project financing.

A2) Considering the national importance placed on the reduction of greenhouse gases by many countries, including the United States, and on economic development and local and regional environmental control by China, our governments should initiate a dialogue on incentive programs to accelerate the deployment of advanced energy technologies, which would become cost-effective in the expected economic environment. Initial support for new and advanced technologies is necessary due to the difficulty in achieving immediate competitive economic results compared to costs of existing deployments. For example, the United States could provide incentives to U.S. companies in the form of tax breaks, low-interest loans, grants, or other mechanisms for undertaking demonstration plants for projects using advanced technologies. Such programs would help to reduce risks to a level acceptable to a private investor. At the same time, the Chinese government could provide incentives such as preferential electricity rates or other means to attract foreign partnerships in advanced energy technologies. The incentives would be provided only for the first few commercial demonstrations of each advanced technology. Thereafter, they would have to compete on a level playing field. Both governments are strongly encouraged to include industry in this dialogue. These types of incentives would be best developed jointly with full transparency on both sides.

A2a)   The CCEF recommends that both governments collaborate on a technology demonstration project to illustrate the mutual benefits of the Clean Development Mechanism (CDM). The CDM would allow United Nations Conference on Parties Annex I nations (including the United States) to earn credits for projects that reduce emissions in non-Annex I countries (including China). Such a demonstration project would set a valuable precedent in U.S.-China collaboration, one that could have a profound impact on the work being undertaken on global change. The initial demonstration project could be structured to be nonbinding for either side and would be intended to be a test of this new mechanism. Financial and other incentives would best be implemented jointly to enhance the effectiveness of such a project.

A2b)   This committee further recommends that the environmental protection agencies of both governments lead a broader governmental collaboration to address environmental degradation associated with fossil fuel burning. This collaboration would focus on better understanding of impacts, most efficient strengthening of emissions standards, creation of clean-energy tax incentives, and other financial and regulatory measures.

A3)   The CCEF recommends a broad participation by agencies from both countries in energy cooperation, with financing agencies and facilities specifically emphasizing their support for energy efficiency, renewable energy, and other advanced clean-energy technologies that would become cost-effective in the expected economic environment. Initial support for new and advanced technologies is necessary due to the difficulty in achieving immediate competitive economic results compared to costs of existing deployments. The committee arrived at this recommendation on the basis of the size and the scale of China's projected growth in the energy sector; the consequent importance of demonstration and deployment of desirable energy technologies; and in the local, regional, and global impacts of the energy sector on health and human well-being.

A3a)   This committee recommends that the U.S. Agency for International Development (USAID) be authorized to include China in its ongoing portfolio of activities. USAID and other key agencies and programs of the U.S. government could do much to assist in many elements of the proposed activities.

Historically, USAID has undertaken activities such as institutional and market reform, technical training, and building and transferring experience with new technologies and management techniques. In particular, USAID works to promote renewable energy applications, energy efficiency, and other clean-energy technologies. These and other USAID functions are needed in China and would complement ongoing technical collaboration through the Department of Energy (DOE), the U.S. Environmental Protection Agency, and the financial vehicles of the Export-Import Bank and the Overseas Private Investment Corporation (OPIC).

The CCEF further recommends that USAID broaden the operating jurisdiction for the U.S.-Asia Environmental Partnership (US-AEP) to include China. US-AEP is a public-private initiative jointly implemented by several U.S. government agencies, under the leadership of USAID. US-AEP currently works with government and industry in eleven Asian countries and is a proven vehicle for environmentally sustainable development. The exclusion of China from such an environmental partnership severely reduces the potential impact of this valuable initiative.

A3b)   The CCEF recommends that the U.S. Trade and Development Agency (TDA) and the Overseas Private Investment Corporation (OPIC) be authorized to conduct activities in China.

About 15 percent of the current TDA budget is used in energy and energy infrastructure projects, and the agency is well positioned to expand these activities to include China. The prefeasibility studies, reverse trade missions, conferences, and technical assistance provided by TDA will aid penetration of established cleaner and more efficient energy technologies and will provide added incentive for U.S. private industry to become involved in the Chinese market. TDA involvement in the Chinese market will support existing initiatives such as the largely underutilized loan program[34] established at the U.S. Export-Import Bank under the Energy and Environment Cooperation Initiative.

OPIC's mission is "complementing the development assistance objectives of the United States" as it helps countries in the transition to market economies, and thereby increases opportunities for U.S. participation in emerging markets. OPIC works in over 140 countries and operates at no net cost to the U.S. taxpayer, as it generates exports and has recorded a positive income each year. In 1997, OPIC reported the potential for $44 billion foreign direct investment for China.[35]

TDA, OPIC, and other U.S. government financial support institutions could help to increase U.S. private-sector participation in the development of China's energy sector by reducing financial risk to acceptable levels. Such modest financial assistance could significantly enhance advanced energy technology utilization opportunities.

B. ENERGY USE AND END-USE EFFICIENCY

There are considerable opportunities for and benefits of strengthening energy efficiency collaboration between China and the United States. China made a deep commitment to energy efficiency in 1980 and has experimented successfully with many different approaches to inculcate efficiency into its energy system over the past two decades. China continues to create and strengthen institutions to promote energy efficiency, and expects to maintain a growth rate of energy demand that is less than half that of its gross domestic product. The United States continues to play a leadership role in the development and dissemination of many energy efficiency technologies. Many of the technologies developed in the United States have been or could be adapted for Chinese markets. Both countries consider energy efficiency to be a cost-effective means of reducing the growth of greenhouse gas emissions as well as important local and regional pollutants, and trade and investment in energy efficiency between the two nations has the potential to yield benefits to both.

Despite of the many interactions underway in the area of energy efficiency (see Chapter 1), there remain high-priority areas for increased collaboration, particularly on:

- incentives to encourage investment and trade in advanced energy efficiency technologies between the two countries, especially important for environmentally beneficial technology;
- cooperative R&D between China and the United States;
- policies and programs to further gains in energy efficiency; and

---

[34] Originally funded at a maximum of $50 million, this program recently has been expanded to a $100 million cap. The first applications for loans under this agreement are being considered by the Export-Import Bank.

[35] From the OPIC website (http://www.opic.gov).

- institutional mechanisms between the two governments.

China has great potential for energy efficiency improvements in major manufacturing industries, power generation, power transmission and distribution, buildings, lighting, appliances, and transportation. A recent assessment concluded that China would be able to save 20 percent of the total current energy consumption by raising its industrial energy efficiency to advanced levels (World Bank, 1997a). China's building standards are designed to cut the heating energy consumption of new centrally heated residential buildings to 30-50 percent of the level of existing buildings. The China Green Lights Program, launched in 1996, is expected to save 22 TWh of electricity, equivalent to 7.2 GW peak-load generation capacity.[36] Of particular urgency is China's need to find solutions to the energy demands and environmental problems caused by fast-growing automobile and other vehicle operations, especially in large cities.

Energy Efficiency Policies and Programs

The recent passage of the Energy Conservation Law in China, combined with the continuing process of reform of energy markets, means that China will be reinventing many of its policy and programmatic approaches to energy efficiency. The mechanisms that China develops in the coming years to enhance markets to energy efficiency (i.e., overcome market barriers) will have a profound effect on the evolution of its energy system. This situation provides a remarkable opportunity for collaboration between China and the United States to have great influence.

The Energy Conservation Law provides only broad guidelines, to be implemented by the State Economic and Trade Commission, the State Development Planning Commission, and regional authorities. Key areas of fruitful collaboration include energy efficiency guidelines and standards for appliances; industrial products; technical guidelines and policy reform for cogeneration; electric utility reform; benchmarking of energy efficiency of new industrial processes; development and enforcement of building energy efficiency codes; supporting new concepts of energy service providers (as distinct from energy providers); and creation and/or modification of institutions to finance energy efficiency.

See cross-cutting recommendations concerning support for joint efforts to assess the impact of energy efficiency policies and programs.

---

[36] The China Green Lights Program is a major energy conservation project in the Ninth Five-Year Plan intended to promote high efficiency lighting products; save electricity; develop a competitive efficient lighting manufacturing industry; and protect the environment. For more information contact the Beijing Energy Conservation Center (BECon).

Investment and Trade in Advanced Energy Efficiency Technologies

The United States has developed many energy-efficient technologies that would be highly beneficial to the Chinese energy system. However, the trade and investment in these technologies has taken place far slower than is desirable for both countries. Some of the barriers include high costs of obtaining reliable information about technologies and markets, inadequately defined business terms (including intellectual property issues), high transaction costs for financing energy efficiency projects, inadequate expertise to evaluate or apply advanced technologies in key energy-using sectors in China, issues of intellectual property agreements in both China and the United States, lack of transparency in terms of commercial trade and investment between the countries, and the numerous barriers to energy efficiency associated with all energy markets. Additionally, there has been very little financial support for activities to broadly promote trade and investment in energy efficiency between the countries, either from private or public sources.

There are compelling reasons at this time to focus increased attention on opportunities to promote commercial transactions in energy efficiency technology between China and the United States. Energy efficiency technologies are a central element of global efforts to reduce the growth of greenhouse gas emissions. The United States as a leading industrial economy and China as a leading developing economy could help to demonstrate the practicality and mutual benefit of bilateral programs to apply advanced technology to promote energy efficiency. Energy efficiency can also be promoted through training programs and control system modifications that enhance efficient operation and maintenance of existing technologies. As noted earlier, it may be possible to use the CDM proposed in climate-change negotiations as a mechanism for promoting Sino-U.S. trade and investment in advanced technology. The potential for energy efficiency technologies to reduce economic costs means that trade and investment could "take off"—yielding substantial business opportunities in both countries—if the barriers to successful business relations can be overcome. This suggests that a relatively modest investment in reducing market barriers may lead to large private-sector opportunities.

B1) The CCEF notes the inadequate support to date for investment and trade in advanced energy efficiency technologies between the two countries and recommends that new resources be devoted to expanding these activities.

B1a)   Governments and international financial institutions should expand existing avenues of information exchange on energy efficiency technologies and policies, support selected training in energy efficiency, and facilitate investment, joint ventures and trade in energy efficiency.

B1b)   The committee recommends that a high-level bilateral special study of institutional innovations to promote financing of energy efficiency be carried out. To be effective, such a study group should have representatives from a variety of public and private financial institutions in order to be able to implement their innovations. Difficulties in establishing mechanisms for financing energy efficiency—because of small size of most projects, high transaction costs, and related market barriers—seriously inhibit technology application.

Research, Development, and Demonstration of Energy Efficiency Technologies

There are many technologies in different stages of R&D for which joint activities between China and the United States could be mutually beneficial.[37] There is a particular need for adaptive R&D, that is, the engineering research necessary to modify a technology for successful application in the circumstances particular to an individual country. Such R&D can, for example, enable the technology to function with low power quality or incompletely processed fuels and simplify its operation or maintenance, integrate technology components into Chinese systems and/or design systems solutions that fit local Chinese conditions and fuel characteristics.

China presently has active demonstration programs on energy efficiency in such diverse areas as lighting, control systems for buildings, utilization of industrial waste, district heating and cooling, and new industrial processes. Both public and private benefits could result from joint demonstration programs.

In some areas of advanced R&D—e.g., the application of information technology to energy efficiency—joint long-term projects composed of Chinese and U.S. researchers could be highly beneficial. Such R&D programs can be stimulated through research exchanges between leading Chinese and U.S. institutions.

B2)   The CCEF recommends significantly strengthening and expanding the existing program of collaborative precompetitive research, development, and demonstration of energy efficiency technologies between the two countries. To date, there has been very little cooperative R&D on energy efficiency between China and the United States. Such R&D can profitably focus on many different areas of energy efficiency in buildings, industry, and transportation. It would be desirable to increase the exchanges between researchers from the two countries, particularly in areas that are highly valued by both countries.

B2a)   The CCEF endorses the activities of the Sino-U.S. Working Group on Energy Efficiency and its subgroups and recommends expanding and strengthening this Working Group as a means of carrying out the initiatives on technology research, development, and demonstration and policy assessment described

---

[37] See the PCAST (1999) report on international cooperation in energy research, development, demonstration, and deployment for a thorough discussion of these issues as they relate to China.

above. Consideration should be given to expansion of the number of teams in the Working Groups to encompass transportation energy efficiency, including work on improving battery and fuel-cell technologies. The move away from petroleum–based transport systems is a logical next step considering our ability to use clean coal technologies, advanced nuclear power and renewable energy sources.

C. CLEAN COAL

The opportunity exists through Sino-U.S. collaborations to deploy a wide variety of clean coal technologies (CCTs) to reduce emissions, improve project performance, and improve the overall system economics of coal use in the United States and in China. This can be done by learning from the U.S. experiences in using some of the technologies, by adapting technologies to suit China's market, and by creating the right market conditions in the United States and China to deploy promising technologies that are not yet used commercially.

Table 3-1 summarizes in broad terms the commercial status of the key CCTs. The United States has had significant experience with most of the technologies in the demonstration and commercial stages, the fruit of a long-term research, development and demonstration program, as well as regulatory requirements and incentives that have created the necessary market conditions for their use. For example, acid-rain control requirements, local siting constraints, and other factors have led to the use a variety of NO_x and SO_2 control devices, configured to meet the requirements of each project. Technologies like coalbed methane recovery (CBM) have been driven both by technology development and regulatory

TABLE 3-1   Commercial Status of U.S. CCTs

Technology	State of Development	Efficiency	Cost	Percent of SO_2 Reduction	Percent of NO_x Reduction
Coal Cleaning	*****	↑	—	20-50	—
Pulverized coal/ scrubber/NO_x	*****	↓	+ +	70-95	20-70
CFBC	*****	—	+ +	50-95	20-50
Supercritical	****	↑↑	+ + +	70-95	20-70
PFBC	***	↑↑	+ + -	70-95	70-95
IGCC	****	↑↑↑	+ + +	80-99	80-98
Vision 21[a]	*	↑↑↑↑	+ + + +	99	90

[a]Vision 21 is a program intended to greatly increase generating and thermal efficiencies with near zero emissions of traditional pollutants and greenhouse gases. For more information contact the office of Fossil Energy, U.S. Department of Energy. See also http://www.fe.doe.gov/coal_power/ fs_vision21.html

TABLE 3-2 Chinese Technical and Economic Evaluation of Different Desulfurization Measures for High-Sulfur Coal

Technical Method	Sulfur Removal Rate (%)	Cost for Removing 1 ton of SO_2 (yuan/t)
Coal cleaning	30-40	500-600
Household briquette	40-50	1500-2000
Industrial briquette	50-70	2000-3000
CFBC desulfurization	85-90	1000-1500
Flue-gas desulfurization	90-95	1700-2200

incentives.[38] Other technologies such as circulating fluidized-bed combustors (CFBCs) have been used extensively in industrial applications because they are economically competitive on the basis of cost penalties for emissions, especially when using low-cost coal and coal wastes.

China also has some commercial experience with CCTs, notably, atmospheric fluidized-bed combustion, pollution control systems, and gasification systems. Some of these experiences have been with advanced CCT technologies, though applications have been limited and usually with the support of government subsidies, not in true market conditions.

In the United States current interest in CCTs is in retrofitting and repowering of existing coal capacity to improve environmental performance. In addition, U.S. industry is interested in developing experience with and reducing the costs of advanced technologies such as pressurized fluidized-bed combustors (PFBCs), integrated gasification combined-cycle (IGCC) in combination with advanced gas turbines and fuel cells, and other technologies that will allow coal to be used more efficiently, cleanly, and economically. However, at least for the foreseeable future in the United States, very little private-sector interest exists to build or invest in research, development, and demonstration on these advanced coal-based systems because of the high risks and long payback of the investments compared to natural gas-based systems.

China, on the other hand, has a near-term interest in using the wide range of CCTs but has difficulty in paying their higher costs or taking the risks associated with them (see Table 3-2 for costs of removing sulfur). In addition, China lacks a strong market pull from environmental regulations and competition. As a result, even with its interest in and need for advanced CCTs, few commercial projects are expected to be undertaken.

The Chinese market is projected to grow rapidly, and retrofitting and repowering could bring substantial benefits. This market therefore would benefit sub-

---

[38] In the United States the Internal Revenue Code "Section 29 production tax credit" legislation gave rise to the development of the CBM industry by effectively providing a higher than market price for gas produced from certain nonconventional sources.

stantially from CCTs, but is not implementing them. This creates a perfect match for the U.S. market that has the CCTs but has no near-term applications for them. Specifically, this situation calls for collaboration between the two countries to adapt existing technologies and create the institutional and regulatory environments needed to attract private-sector investment in these technologies in China. In return U.S. industry would benefit by learning about the cost-cutting advantages of deploying these technologies.

Planning and implementing such an initiative to meet these objectives has several parts: (1) focused information exchange and dissemination, (2) research to adapt the technologies for the Chinese market, (3) institutional and regulatory reform, (4) practical application of CCTs in China, and (5) deployment and transfer of the experience and technologies back to the United States and other coal-burning economies.

Sino-U.S. collaborations on CCTs have been under way since 1979. These longstanding collaborations have transferred information, conducted studies, provided training, undertaken educational programs, and instituted a variety of activities intended to promote use of these technologies in China. However, despite the importance of coal to both China and the United States, there are few well-defined and long-term continuing initiatives. Dramatic changes in the Chinese and U.S. governments as well as in industry have changed policies, priorities, and personnel. As a result, a new mechanism for prioritizing and coordinating CCT studies, research, and technology transfer is required if CCTs are to play a significant role in the short and mid terms. The four Academies are in a good position to play a role this activity; specific roles for the Academies can be found in Recommendation A1.

Information Exchange and Dissemination

C1) The CCEF recommends that the U.S. Department of Energy and the Chinese Ministry of Science and Technology review and strengthen the programs and processes for utilizing and disseminating CCTs in China. This will help to define what more needs to be done, how it should be done, and by whom. Two general components of the effort should be addressed: (a) the need for an ongoing, widely recognized central information repository where existing data on CCTs can be kept and to which new information could be added and (b) a series of focused workshops on carefully chosen top-priority CCT topics. A significant shortfall of government efforts to date is the limited participation of private industry and the narrow focus on a specific technology.

C1a) The CCEF recommends that both governments support a non-governmental, independent Clean Coal Technologies Information and New Technology Training Center, which would cooperate closely with the private sector in each country to undertake studies and analyses, conduct seminars, and perform institution building. The focus of this recommendation is information collaboration, not joint R&D. Some of these information activities in the CCT area are already underway—especially project-specific opportunities—but such efforts would greatly benefit from an international institutional entity that would act as a clearinghouse for CCT information. An important component of an information dissemination program would be to stress the importance of rational coal pricing schemes and recent successes in restructuring the coal industry.

C1b) The committee identified several key areas in which workshops and training might promote increased application of CCTs in China. These include use of modern management tools and procedures to optimize plant performance, examination of progress at various CCT demonstration projects worldwide, and increased analysis and discussion of how international financial institutions can contribute to the application of CCT projects in developing countries.

CCT Adaptation Research and Technology Demonstration

Many CCTs that are in use throughout the world are not being given serious consideration in China. In most cases, this is because they have been designed for markets in developed countries that have stringent environmental requirements, market conditions, or coal characteristics that differ significantly from those in China. These technologies may play an important role in China's energy future, but only after adaptation to local energy needs, feedstocks, and market conditions. For example, many power plants in China are reluctant to use washed coal because it adds to fuel costs despite the fact that using cleaner coal could reduce operation and maintenance costs.

C2) The CCEF recommends that both governments convene a group representing public and private interests to assess the variety of CCTs, to determine their suitability for China's market and to identify adaptations that will be required for each technology to make it more suitable for near-term use. This group would create a detailed research plan and suggest appropriate funding levels and sources (not limited to governmental support) in a broad implementation plan. This group would not consider development of entirely new technologies; rather it would aim to identify ways to adapt existing advanced technologies to accelerate near-term use and to encourage continued technology improvement for long-term sustainability.

The key to the success of this activity will be the inclusion of industry representatives (such as the Electric Power Research Institute in the United States, and a variety of technical institutions in China) to build upon the extensive foundation of existing technical studies and other initiatives listed in Chapter 1. Close coordination with the Clean Coal Technologies Information and New Technology Training Center mentioned earlier would be desirable to reach as broad a range of candidate technologies as possible. The objective of this effort would be to identify the CCTs that could be most important to China considering economic, environmental and energy factors and to define the research that needs to be done to ready these technologies for the Chinese market.

C2a)   On the basis of preliminary analysis of priority areas for Chinese application of CCTs, the committee suggests the following high-priority areas for cooperative research, development, demonstration, and deployment in the mid term:

- *coal-fired combined-cycles such as IGCC;*
- *increased efficiency and cleaner combustion of low-heating-value fuels;*
- *lower-cost SO_2 control technologies (addressing both the manufacturing costs for flue-gas desulfurization equipment and plant operational costs);*
- *cleaner combustion, desulfurization, and dust removal in small and medium-size coal-fired boilers; and*
- *techniques to maximize the contribution of CBM while reducing environmental and safety hazards.*

Because of the large variety of coal, some of which is of poor quality, circulating fluidized bed technology will be of great interest to China. For the intermediate term, the hybrid CFB-based combined-cycle holds promise, especially if these can be retrofitted to the fleet of existing CFBs to increase capacity and efficiency. Advanced PFBC and IGCC will be pursued in the near term only in specific limited applications under specific market conditions.

The committee also identified several issues specific to development of CBM resources that bear greater scrutiny; these are covered in Recommendation D1 and in Box 1-1.

D.   NATURAL GAS

The natural gas sectors of the United States and China are very different in their scale and technical development, though each country will be relying more on gas resources in the time frame of this study. Cooperation between the two countries would bring new business opportunities to the United States and capital, technology and business knowledge to China.

Historically, natural gas has lagged in China's energy development, although recently growing concerns about severe air pollution in urban areas have made natural gas a high priority for displacing urban coal consumption. Natural gas also could play an important role in reducing carbon dioxide emissions in the short to medium term if leak-tight technologies are employed. In the longer term, gas from various sources figures prominently in all scenarios for successfully and cleanly meeting energy needs. Some of the advantages of natural gas include lower capital costs, decreased emissions, shorter plant construction times, modular and scalable units, and increased reliability compared to coal. Consideration of joint initiatives thus should pay greater attention to the issues and challenges in increasing the demand and supply of natural gas.

For natural gas to have a significant role in China's energy future, a great deal more natural gas is needed, and for environmental reasons, needed sooner rather than later. There is little doubt that China's natural gas resources would be able to support a demand much larger than the current level. China also may choose to import natural gas to supplement its domestic production. There are high hopes for much greater use of natural gas in China, but the challenges are many and large.

On the supply side, China needs to locate more ready-for-development natural gas reserves, and to accelerate their development. The transmission, distribution, and storage infrastructure has to be greatly expanded to ensure that produced and imported gas reaches the market. This would require huge capital investments as well as acquisitions of a great amount of technology and business knowledge in a relatively short period of time. China would need foreign participation to meet these challenges, however, the current natural gas sector policy framework is not conducive to significant outside participation. Major obstacles include:

- a lack of clear rules, regulations, and consistent approval procedures for foreign investments in the oil and gas sector;
- the lack of access to high-quality exploration areas and crucial data and information;
- restrictions on foreign equity investment in pipelines; and
- gas prices that are artificially controlled below their market clearing levels, creating difficulties for both domestic and foreign investors in recovering their costs.

D1) The CCEF recommends that both governments work collaboratively to explore possibilities in developing an overall strategy for accelerated natural gas development in China that includes production of domestic natural gas and CBM, and imports of piped natural gas and liquefied natural gas. Such a strategy also would address the associated infrastructure improvements—transportation, distribution, and use—which represent many of the challenges and investment costs of the natural gas industry. The policy framework for this development needs to be market driven, with transparent incentives and regulations. As a world leader in natural gas and CBM development, the United States could be a very useful partner for China in this arena.

D1a) The committee wishes to highlight specific portions of this overall strategy that would benefit from increased near-term technical and institutional collaboration (some of these issues are addressed in a more general context in the cross-cutting recommendations above):

- *assisting in exploration and resource assessments for the varying geological conditions of Chinese gas reserves;*

- *developing a market infrastructure, including physical facilities and institutional arrangements for an efficient and reliable gas supply system;*
- *undertaking further environmental policy reform to better allow natural gas to compete with coal;*
- *developing high-priority areas specific to CBM, including techniques for exploration, drilling, testing, evaluation, and reservoir protection of CBM resources, as they differ from natural gas deposits (see also Box 1-1).*

D1b)   The CCEF strongly emphasizes the importance of CBM to China's efforts to diversify its future energy supplies. Extracting CBM increases mining safety, reduces greenhouse gas emissions, and provides a low-pollution energy supply. The benefits are substantial, but to bring CBM to large-scale commercial utilization, China must deal with development problems associated with natural gas in general, as well as specific issues concerning CBM. Exploration and resources assessment for CBM are at a very early stage in China. CBM reservoirs tend to be fragmented and smaller than those of conventional natural gas, raising the costs of development. In this regard, successful CBM policies need to be more creative and incentive-laden than those for natural gas. China has made great progress in development of CBM resources thus far and is well positioned to continue this trend.

Such an upgrading of China's technological capacity to assess, develop, and efficiently utilize CBM would be a valuable component of broader policy and regulatory initiatives aimed at fostering the development of a complete CBM industry and the necessary infrastructure to support it. Neglect of these considerations leading to the release of the large amount of methane could seriously compound greenhouse gas problems.

The committee also notes that increased activity in the coalbed methane industry in China could speed the development of China's natural gas industry because the two industries share many of the same techniques and infrastructure, both physical and institutional. Also, China is gaining valuable experience experimenting with financial incentives and more flexible contracts for development of CBM (e.g., tax holidays and reduced royalties, as well as freedom to select individual exploration areas, as opposed to preselected lots), which also can be transferred to a burgeoning natural gas industry.

Cooperative research, development, and demonstration in several key technology areas would help to ensure the rapid development of China's CBM industry. The first set of CBM projects undertaken in China faced difficult geological conditions and were hampered by the inexperience of the Chinese oilfield services industry in adopting CBM drilling and testing techniques (Stevens, 1999). These are outlined in Box 1-1.

D1c)   In developing an overall natural gas utilization strategy, the CCEF recommends China consider distributed electric power generation options using remote sources of natural gas or CBM from smaller fields to meet the energy needs of remote populations currently without access to commercial energy and to augment existing services through increased reliability and lower total cost. Distributed generation might prove to be less of a challenge in the short term than embarking on large-scale natural gas infrastructure projects. Should further exploration reveal sizable natural gas resources, infrastructure could be developed to add natural gas to the coal, nuclear, and hydro that already supply electricity to large urban areas.

E. PETROLEUM

The petroleum sectors of China and the United States also differ greatly in scale, technical development, competitiveness, and openness to the international market; and each country has different priorities. Each country, however, is facing an increasing reliance on imported petroleum products and shares common concerns for energy security (see "Import Dependence and Energy Security" in Chapter 2).

In China the costs of crude oil production have increased rapidly in the past decade because of efforts to maintain output from the dwindling pools of held reserves. Inefficient management and technology constraints also have contributed to reduced productivity of many onshore oil fields. Greater exploration efforts are needed to materialize the promises of western China, and huge investments will be required to develop those oil resources and to bring them to the eastern consumption centers.

Increasing demand for imported crude oil, especially that from the Gulf region, will put great pressure for China's refining industry to upgrade and adjust production and pollution control facilities. Most of the existing refineries were designed to handle low-sulfur and relatively heavy Chinese crude and will not be able to easily process the high-sulfur Gulf crude. The refining industry has begun the process of phasing out leaded gasoline production, and this process needs to be accelerated for important public health reasons.

E1) The CCEF endorses the objectives of the ongoing "Oil and Gas Forum" whose activities are outlined in Chapter 1 and recommends that the following major areas for cooperation in the petroleum sector between China and the United States be on the agenda of this continuing bilateral dialogue:

- petroleum sector restructuring, focusing on building market institutions, reforming corporate/enterprise management, and improving customer services;
- long-term sector strategies on indigenous oil development, international operations, oil imports, and strategic reserves;
- exploration and resources assessment;
- refining technologies, phasing out of leaded gasoline, and alternative transport fuels; and

- environmental protection in the oil development, production, and consumption chain, in a system that reflects real costs.

E2) The CCEF found common energy security concerns that stem from each country becoming more dependent on petroleum imports and recommends two action items:

E2a)   China and the United States should collaborate in a comprehensive analysis of the potential and merits of national and regional strategic petroleum reserve systems. Development of a strategic petroleum reserve for China could ease concerns over energy security and also help to insulate China from temporary price fluctuations in the international oil market. In collaborating to examine the needs for and experience with the U.S. strategic petroleum reserve, both countries stand to gain. Further, both governments are strongly encouraged to work closely with the International Energy Agency on this topic.

E2b)   Both governments should collaborate in a strategic study of the macroeconomic impact of fluctuations in the world petroleum market, and institutions and measures to minimize these disturbances, including the influence of oil future and spot markets.

E3) The CCEF recommends that the U.S. and Chinese governments and industry establish a dialogue on light transport vehicles, including alternatives to petroleum transport fuels, and cooperate on both technology development and market creation. This dialogue would identify specific collaborative opportunities for more detailed work in coal-based multigeneration systems, coal and biomass liquefaction, fuel cells, and battery technologies.

E3a)   The committee also noted the need for a broader examination of urban transportation systems in China and the United States. Alternative options to provide necessary mobility could have profound consequences for the energy sector, especially in petroleum use. The Organisation for Economic Cooperation and Development (OECD) has undertaken efforts to examine transportation systems in light of sustainability concerns, and could provide valuable insight.[39]

F.   RENEWABLE ENERGY

The Chinese government's strategy for greater utilization of renewable energy technologies is outlined in the New and Renewable Energy Development Program, 1996-2010, jointly issued by the State Economic and Trade Commission, the State Planning Commission, and the State Science and Technology Commission. To ensure the realization of the program goals and to make renewables

---

[39] See the work undertaken by the OECD Environment Directorate project on Environmentally Sustainable Transport (http://www.oecd.org/env/trans/) and earlier work by the OECD/European Council of Transport Ministers Working Group on Urban Travel and Sustainable Development.

a commercially viable energy source in the long term, China must overcome the technical and institutional barriers, discussed in previous chapters, through a strong and steady effort. With timely and targeted international assistance, this process could be accelerated and the costs could be reduced.

The U.S. experience—much of it of limited success—has shown that for renewable energy technologies to become self-sustaining in a market economy, targeted government assistance in developing market infrastructure, improving information access, increasing commercial capabilities, and supporting R&D is crucial. Using market-based incentives where and when necessary, such as the time-restricted tax credit for wind-generated electricity in the United States, is often an effective way to attract investment and speed commercialization. Government policies should strive to increase the demand for promising renewable energy technologies and to clarify environmental objectives.

The importance of collaboration is clear: Although developed nations currently dominate in experience and development of renewable energy resources, the longer-term market for these energy technologies will largely be in developing countries (WEC/IIASA, 1995). The issues and obstacles surrounding technology transfer are not trivial: Widespread deployment of renewable technologies will require significant technological adaptation to local needs and uses. The key to the deployment of renewable energy technologies in any significant capacity is collaboration.

F1) The CCEF finds that U.S.-Chinese cooperation in the following renewable energy areas would be especially helpful:

• setting up a market-oriented policy framework for sustained expansion of renewable energy applications, especially in the areas of financial incentives, credit policy, market regulations, and industry standards;
• determining commercial priorities through technology and market assessments;
• strengthening R&D cooperation and trade and investment in advanced renewable energy technologies, including scholarly exchanges, technology demonstrations, and joint ventures; and
• training renewable energy practitioners and entrepreneurs.

The committee also notes that USAID, TDA, and OPIC have a great deal of experience in the area of renewable energy technologies and cleaner energy systems, and could work within established programs to further the goals listed earlier. Cooperation in these areas would be of significant mutual benefit. (See also Recommendation A3.)

F1a) To ensure that these large-scale renewable energy technologies are available for widespread deployment in the middle of the next century, the CCEF recommends that the governments of the United States and China consider a long-term R&D public-private partnership program taking advantage of the strengths of both countries' institutions. This partnership should consider that the most important approach to increase the use of renewable energy in both China and the United States will be to lower the cost per unit of energy delivered, and this should be the first priority of a collaborative program.

F1b)   The committee found that collaboration, trade, and investment in the following advanced renewable energy technologies may have significance in the time frame of this study:

- *design and implementation of large-scale, grid-connected wind farms;*
- *advancement of the solar photovoltaic (PV) and solar thermal collector industries, with a goal of improving quality, increasing efficiency, and, most importantly, lowering costs;*
- *demonstration of surplus biomass cogeneration for grid-connected plants;*
- *identification of promising feedstocks and applications for biomass gasification;*
- *technical assistance in the commercialization of biomass gasification technologies;*
- *development of anaerobic fermentation technologies in power generation and wastewater treatment; and*
- *hydroelectric power projects for China.*

The committee found that the focus of collaborative efforts for commercial and near-commercial technologies—such as solar water heaters, small PV, and wind power—should be on lowering costs of production, which is primarily an industrial responsibility. For demonstration technologies—such as solar thermal (both dish and trough), rooftop PVs, biomass, and municipal waste technologies—the focus should be to organize projects in China to introduce their advantages to potential implementing institutions and to reduce costs through scale-up and experience. Perhaps the most significant cooperation in the renewable arena will be in pre-competitive cooperative R&D in China and the United States in areas such as new PV cells, hydrogen energy, and storage components in renewable systems: These technologies represent the eventual transition away from fossil fuels and are of considerable importance outside of the time frame of this study. Research on these and other advanced renewable technologies are sometimes beyond the scope of commercial R&D and might benefit from government-supported activities.

F2)   The CCEF recommends that both governments establish periodic reviews of renewable energy collaboration to better meet strategic objectives of both countries. Renewable energy programs are being carried out by different government agencies, international organizations, and research institutes. A regular review of the progress of renewable energy collaborations—including solar, wind, biomass, and hydro—and frequent information exchange among institutions would result in better coordination of international programs

G. NUCLEAR ENERGY

The near-term focus in the United States is on keeping nuclear plants currently in existence operating safely to the end of their current 40-year license periods, and, if the economics of these plants are deemed competitive, for an additional 20-year period through a relicensing process. To a country such as China, which is just now embarking on a commercial nuclear power generation program, the economics of nuclear power improve if plants can be shown to operate safely beyond their original design life.

China and several other Asian nations view nuclear power as a necessary option for ensuring energy security and environmental improvement. Globally, nuclear power must be viewed as an option having potential significance in both a sustainability and an environmental context in the first half of the next century. Operational performance and economics of nuclear power plants continue to improve, with International Atomic Energy Agency (IAEA) and the World Association of Nuclear Operators (WANO) both playing important roles. Advanced light water reactor (ALWR) designs have been introduced for both pressurized water reactors and boiling water reactor technologies to further enhance the safety of future plants. Other designs, such as gas cooled reactors, also show promise for enhancing safety of future nuclear plants.

These positive developments over recent years notwithstanding, there are no prospective orders in the next several years in the United States or Europe and few in Japan. In the United States, orders for natural gas power plants will dominate in the next several years given the combined effect of lower capital costs, relatively low-priced fuel, and a highly efficient combined-cycle technology. Should greenhouse gas controls be imposed, though, the attractiveness of nuclear power would rise, and the option to expand its use should be maintained.[40]

G1) The CCEF found that the following priorities for our governments concerning commercial nuclear power programs are very similar: the ability to confidently prevent proliferation of fissile materials and handle spent fuel and waste; safety in the design and operation of nuclear plants; and a desire to improve the economics of nuclear plants. Without resolution of these three issues and, in the United States, increased public acceptance, the future of nuclear fission remains in doubt. Given the need to resolve these issues this committee articulated some fundamental principles in our common interest.

---

[40] Nuclear power becomes economically competitive with fossil fuels and renewable energy sources in an Energy Information Administration policy case in which emissions must be reduced 3.5 percent from 1990 levels.

G1a)   The committee suggests that to minimize future risk and costs an important consideration is to simplify and standardize the design of future plants. This is a lesson learned from the U.S. and European experience with nuclear power plant operations.[41] This committee saw an opportunity for collaboration on the joint design and construction of a 1000-MW ALWR specifically intended for the Chinese market. This plant would be first-of-a-kind project and would strive for simplicity and standardization to improve the economics as well as safety in operation.

G1b)   The committee found that the successful demonstration and acceptance of a long-term disposal and storage option that would handle either spent fuel or waste is needed for a continued commercial nuclear power program. This would remove one of the uncertainties or challenges of nuclear power without prejudging what the future choices will be—based on security, economic, and environmental considerations—on the closure of the fuel cycle.

G2)   Our governments and industry should play a leadership role in international organizations such as the International Atomic Energy Agency and the World Association of Nuclear Operators to assure that international commitments, regulations, and appropriate measures are defined and implemented. The IAEA International Nuclear Safety Advisory Group, the Institute for Nuclear Power Operations, the OECD Nuclear Energy Agency, and the Nuclear Suppliers Group would benefit from strong leadership. The committee also notes the importance of the role played by these organizations in efforts to improve the quality of information available to the public with regard to nuclear power. Such an effort should present a balanced discussion of the costs and benefits—notably environmental—associated with nuclear fission.

G3)   The committee emphasizes the importance of bilateral cooperation and endorses the framework of the Agreement on Intent of Cooperation Concerning Peaceful Uses of Nuclear Technology (PUNT) signed by both governments in 1997. The committee encourages Chinese scientists and regulators to use this framework agreement to establish a formal relationship with counterparts in the United States to address technical issues, to exchange views on policy and regulatory controls in commercial nuclear power, and to coordinate the development and training of personnel for regulation of nuclear power in China and the United States. This initiative requires strong leadership and commitment from within government, but also should encourage the active participation of the private sector in exploring these issues.

G3a)   The CCEF recommends that the U.S. DOE consider expanding the Nuclear Energy Research Initiative (NERI) to cooperative efforts involving China

---

[41] In the United States there are 80 different plant designs from 4 different vendors.

and other countries, as appropriate. The U.S. and China share many long-term goals and are each working with the same set of technologies and issues. Such a cooperative approach would make the best use of each country's R&D budget, and would allow tailoring of technologies and techniques to the specific conditions in which they are expected to be deployed.

G3b)   To address vocation training, the committee recommends that each country consider the model and practices of the National Academy for Nuclear Training to further the goal of ensuring high-quality personnel and safety standards at each country's nuclear facilities.[42] Universities should be encouraged to promote interest for, and expand university level courses in, nuclear engineering to ensure a steady pool of scientific talent to staff the nuclear energy industry. The Chinese government has also specifically articulated the goal of increasing education and training in nuclear science and engineering, and these remain high priority items.

H.   ELECTRICITY TRANSMISSION AND DISTRIBUTION SYSTEMS

Both China and the United States are moving to cleaner, more efficient energy systems, and electricity is the most appealing choice. China's electricity use per capita is about half of that of Brazil and one-fifteenth that of the United States. Electrification is an urgent priority in China for economic, social, health, and environmental reasons: Energy intensity in many countries has decreased with increased electrification; and the harmful effects of direct coal use—particularly the significant environmental impacts—can be avoided through increased electrification. Electrification also can provide commercial energy to unserved populations.

H1) The CCEF recommends that the governments of the United States and China collaborate on measures to foster the development of a successful electric power sector, including:

*   • *planning for interconnection and further development of the electric power grid*
*   • *support for and/or initiation of financing by the World Bank, Asian Development Bank and other international financial institution for the electric power grid in China;*

---

[42] The National Academy for Nuclear Training (NANT) was formed under the auspices of INPO in 1985 to focus industry efforts on a nationwide basis to continue improvements in training and promote professionalism of nuclear plant personnel. NANT integrates the training resources and activities of all U.S. nuclear utilities and an independent 20-member accrediting board.

• *study of the legal limitations of foreign participation in, and financial support for, electric power transmission facilities in China to promote increased interest in independent power production in China; and*

• *exploration of options to improve the adequacy, quality, and reliability of electric power and the reduction of line losses in transmission and distribution.*

China is facing a difficult task in the interconnection of the six regional power networks and could benefit greatly from collaboration with institutions in the United States. Specific technology expertise is needed, which includes many aspects of establishing and maintaining a reliable interconnected power grid, such as techniques for load leveling and load profile improvement, power quality, and the upgrading of distribution systems.

A structured exchange between the Electric Power Research Institutes in each country would provide significant opportunities for information exchange, and could provide insight into advanced technology deployment in the United States especially in flexible AC transmission (FACTs), performance of clean coal technologies, and distributed generation deployment. Such a relationship could also provide the connection necessary to build on the experience gained in the United States in providing power to outlying rural areas.

The CCEF recommends that both governments consider initiating a project that demonstrates the potential for programs in demand-side management, load management, and integrated resource planning, starting from a market-based approach to energy pricing. The committee noted the wealth of expertise in the U.S. utility and power services industries on these techniques, which, if adapted to local conditions, could play a significant role in China. The United States is moving away from these activities as it deregulates its electric power industry; the timetable for China's eventual move to deregulation, however, is unknown, and that decision will be based on its experience in several trial areas.

H2) The CCEF also found that China and the United States share an interest in developing more economically viable distributed power sources for remote areas, as noted in the recommendation on natural gas. China's ongoing efforts to provide energy services to its large rural population provide a significant opportunity to examine the role of non grid-connected systems, especially those that incorporate a renewable energy component. Efforts of international financial institutions to provide electric power services in China are encouraged to look at opportunities in distributed generation in China.

Acronyms

ALWR	advanced light water reactor
AFBC	atmospheric fluidized bed combustion
APEC	Asia Pacific Economic Cooperation forum
BWR	boiling water reactor
CAE	Chinese Academy of Engineering
CANDU	Canada Deuterium Uranium
CAS	Chinese Academy of Sciences
CBM	coalbed methane
CCT	clean coal technology
CFBC	circulating fluidized bed combustion
CNG	compressed natural gas
CO	carbon monoxide
CO_2	carbon dioxide
CUCBM	China United Coalbed Methane Corporation
DOE	U.S. Department of Energy
DSM	demand side management
EIA	U.S. Energy Information Administration
EPA	U.S. Environmental Protection Agency
ESCO	energy services company
FACT	flexible alternating current transmission
FERC	Federal Energy Regulatory Commission

FGD   flue gas desulfurization

GDP   gross domestic product
GEF   Global Environment Facility
GHG   greenhouse gas

IAEA   International Atomic Energy Agency
IFI   international financial institution
IGCC   integrated gasification combined cycle
IIASA   International Institute for Applied Systems Analysis
INSAG   International Nuclear Safety Advisory Group
IOM   Institute of Medicine
IRP   integrated resource planning

LNG   liquified natural gas
LPG   liquified petroleum gas

MDB   multilateral development bank
MOX   mixed oxide fuel

NAE   National Academy of Engineering
NAS   National Academy of Sciences
NEPA   Chinese National Environmental Protection Agency
NNSA   Chinese National Nuclear Safety Administration
NO_x nitrogen oxide
NRC   U.S. Nuclear Regulatory Commission

OECD   Organisation for Economic Development and Cooperation
OPEC   Organization of Petroleum Exporting Countries
OPIC   Overseas Private Investment Corporation

PCAST   U.S. President's Council of Advisors on Science and Technology
PFBC   pressurized fluidized bed combustion
PUNT   Agreement on Intent of Cooperation Concerning Peaceful Uses of
             Nuclear Technology
PWR   pressurized water reactor
PV   photovoltaic

R&D   research and development
RD&D   research, development and demonstration
RPS   Renewable Portfolio Standard

SDPC   Chinese State Development Planning Commission

SETC	Chinese State Economic and Trade Commission
SO_2	sulfur dioxide
SSTC	Chinese State Science and Technology Commission
T&D	transmission and distribution
TDA	U.S. Trade and Development Agency
TSP	total suspended particulates
UNDP	United Nations Development Programme
USAID	U.S. Agency for International Development
WANO	World Association of Nuclear Operators
WEC	World Energy Council

UNITS OF MEASURE

b/d	barrels per day
billion m^3	billion cubic meters
Btu	British thermal unit
dollar	one U.S dollar equaled 8.28 renminbi in 1998
EJ	exajoule
GW	gigawatt
km	kilometer
kw	kilowatt
kWh	kilowatt hour
m^2	square meters
m^3	cubic meters
mm	millimeter
mt	metric tons
mtce	million tons of coal equivalent
MW	megawatt
MWp	megawatts peak power
Quad	quadrillion Btu
RMB	renminbi or yuan
tcf	trillion cubic feet
TWh	terawatt hours

Energy Conversion

ENERGY CONVERSION FACTORS

From one:	To:	EJ	Btce	Btoe	Tcm NG	Quad
Exajoule	*EJ*	1.000	0.033	0.022	0.025	0.948
Billion metric tons coal equivalent [2]	*Btce*	30.300	1.000	0.675	0.761	28.720
Billion metric tons oil equivalent [3]	*Btoe*	44.900	1.482	1.000	1.128	42.559
Trillion cubic meters natural gas [4]	*Tcm NG*	39.800	1.314	0.886	1.000	37.725
Quadrillion Btu	*Quad*	1.055	0.035	0.023	0.027	1.000

[1] These factors are as per footnote 1 on p. 2, and follow the U.S. convention of high-heat values.
[2] Chinese conversion factors for coal and other fuels are low-heat values. Chinese average raw coal contains 20.93 GJ/metric ton (low heat), or 22.51 GJ/t (high heat), assuming that low-heat values for coal are 93% of high-heat values. China typically converts all its energy statistics into "metric tons of standard coal equivalent" (tce), a unit that bears little relation to the heating value of coals actually in use in China. One tce equal 29.31 GJ (low heat) equivalent to 31.52 GJ/tce (high heat).
[3] China uses a conversion factor for its oil of 41.87 GJ/metric ton (low heat), equivalent to 44.07 GJ/t (high heat), assuming that low-heat values for oil are 95% of high-heat values.
[4] China uses a conversion factor for its natural gas of 38.98 GJ/thousand cubic meters (low heat), equivalent to 43.31 GJ/tcm (high heat), assuming that low-heat values for natural gas are 90% of high-heat values.

Chinese factors:		low heat	conversion	high heat
average raw coal	GJ/t	20.93	0.93	22.51
standard coal equivalent	GJ/tce	29.31	0.93	31.52
oil	GJ/t	41.87	0.95	44.07
natural gas	GJ/tcm	38.98	0.90	43.31

Both tables courtesy of J.E. Sinton, Energy Analysis Division, Lawrence Berkeley National Laboratory.

Multiplier Prefixes for use with International System of Units (SI) Prefixes

kilo = thousand = 10^3
mega = million = 10^6
giga = billion = 10^9
tera = trillion = 10^{12}
peta = quadrillion = 10^{15}
exa = quantrillion = 10^{18}

Adapted from EIA, Annual Review of Energy 1997

Abbreviations

Quad = quadrillion (10^{15}) British thermal units (Btu)
mtce = million ton of coal equivalent
mtoe = million ton of oil equivalent
boe = barrel of oil equivalent

One barrel of oil = 0.136 tons of oil
One short ton (2000 lbs.) = 0.907 metric tons
One cubic foot = 0.0283 cubic meters

References

BMI/BECC/ERI (Battelle Memorial Institute, Beijing Energy Conservation Center, and the Energy Research Institute of China). 1998. *China's Least Cost Power Options: An Analysis of Economic and Environmental Costs.* Washington, D.C.: Battelle Memorial Institute.

Cai Ruixian. 1998. *Energy in China.* Institute of Engineering Thermophysics, Chinese Academy of Sciences. (mimeo)

Cai Ruixian and Cui Zhengxin. 1998. *Chinese Energy Situation.* Institute of Engineering Thermophysics, Chinese Academy of Sciences. (mimeo)

DOE (U.S. Department of Energy) and CAS (Chinese Academy of Sciences). 1996. *U.S.-China Experts Report on Integrated Gasification Combined Cycle Technology.* Washington, D.C.: U.S. Department of Energy.

EIA (Energy Information Administration). 1997. *Natural Gas Annual.* Washington, D.C.: U.S. Department of Energy.

EIA (Energy Information Administration). 1998a. *Natural Gas 1998: Issues and Trends.* Office of Oil and Gas, Energy Information Administration, U.S. Department of Energy. Washington, D.C.: U.S. Department of Energy.

EIA (Energy Information Administration). 1998b. *U.S. Crude Oil, Natural Gas, and Natural Gas Liquids Reserves: 1997 Annual Report.* Washington, D.C.: U.S. Department of Energy.

EIA (Energy Information Administration). 1999. *Annual Energy Outlook.* Washington, D.C.: U.S. Department of Energy.

EPRI (Electric Power Research Institute) and DOE (U.S. Department of Energy). 1997. *Renewable Energy Technology Characterizations.* TR-109496. Palo Alto, CA and Washington, D.C.: EPRI and U.S. Department of Energy.

Fan Weitang, Cheng Yuqi, and Pan Huizheng. 1998. *Chinese Clean Coal Technology.* Chinese Academy of Engineering. (mimeo)

Fan Weitang and Sun Maoyuan. 1998. *Present Situation and Prospects of Oil, Natural Gas and Coalbed Methane Industries in China.* Chinese Academy of Engineering. (mimeo)

Garbaccio, R., M. Ho, and D. Jorgenson. 1998. *Controlling Carbon Emissions in China.* Cambridge, MA: Harvard University.

Gas Research Institute. 1999. *U.S. Coalbed Methane Resources.* Chicago, IL: Gas Research Institute.

Hu Jianyi. 1998. *Fossil Fuel in China.* Research Institute of Petroleum Exploration and Development, China National Petroleum Corporation. (mimeo)

May, Michael. 1998. *Energy and Security in East Asia.* Palo Alto, CA: Asia/Pacific Research Center, Institute for International Studies, Stanford University.

Nakicenovic, N., A. Grübler, and A. McDonald, eds. 1998. *Global Energy Perspectives.* Cambridge, England: Cambridge University Press.

National Research Council. 1992. *The National Energy Modeling System.* Washington, D.C.: National Academy Press.

Office of Economic, Electricity and Natural Gas Analysis, Office of Policy, U.S. Department of Energy. 1999. *Supporting Analysis for the Comprehensive Electricity Competition Act* (DOE/PO-0059). Washington, D.C.: U.S. Department of Energy.

ORNL (Oak Ridge National Laboratory) and RFF (Resources for the Future). 1992-96. *External Costs and Benefits of Fuel Cycles: A Study by the U.S. Department of Energy and the Commission of the European Communities (ORNL/M-2500).* Washington, D.C.: McGraw Hill/Utility Data Institute.

Report Number 1   *Background Document to the Approach and Issues.* 1992.
Report Number 2   *Estimating Fuel Cycle Externalities: Analytical Methods and Issues.* 1994.
Report Number 3   *Estimating Externalities of Coal Fuel Cycles.* 1994.
Report Number 4   *Estimating Externalities of Natural Gas Fuel Cycles.* 1996.
Report Number 5   *Estimating Externalities of Oil Fuel Cycles.* 1996.
Report Number 6   *Estimating Externalities of Hydro Fuel Cycles.* 1994.
Report Number 7   *Estimating Externalities of Biomass Fuel Cycles.* 1996.
Report Number 8   *Estimating Externalities of Nuclear Fuel Cycles.* 1995.

PCAST (President's Council of Advisors on Science and Technology, Panel on International Cooperation on Energy Research, Development, Demonstration, and Deployment). 1999. *Powerful Partnerships: The Federal Role in International Cooperation on Energy Innovation.* Washington, D.C.: PCAST.

SETC (State Economic and Trade Commission). 1997. *China Energy Annual Review.* Beijing, China: State Economic and Trade Commission.

Simpson, R. David, ed. 1999. *Productivity in Natural Resource Industries,* Washington, D.C.: Resources for the Future.

Sinton, Jonathan E., ed. 1996. *China Energy Databook.* Berkeley, CA: Ernest Orlando Lawrence Berkeley National Laboratory.

Smil, Vaclav, and Mao Yushi (coordinators). 1998. *The Economic Costs of China's Environmental Degradation.* Cambridge, MA: American Academy of Arts and Sciences.

Stevens, Scott H. 1999. China coalbed methane reaches turning point. *Oil and Gas Journal,* January 25. pp 101-6.

U.S.-China Oil and Gas Industry Forum. 1997. U.S.-China Energy and Environment Cooperation Initiative Fact Sheet. October.

U.S. National Laboratory Directors. 1997. *Technology Opportunities to Reduce Greenhouse Gas Emissions.* Washington, D.C.: U.S. Department of Energy.

U.S. National Laboratory Directors. 1998. *Scenarios of U.S. Carbon Reductions: Potential Impacts of Energy-Efficient and Low-Carbon Technologies by 2010 and Beyond.* Washington, D.C.: U.S. Department of Energy.

Wang Yangzu and Lu Xinyuan. 1997. *Study in Design and Implementation of Pollution Levy System in China.* Beijing, China: Chinese Research Academy of Environmental Sciences.

Wang Yingshi. 1998. *Pollution and Environmental Issues in China due to Coal Burning.* Institute of Engineering Thermophysics, Chinese Academy of Sciences. (mimeo)

WEC/IIASA (World Energy Council/International Institute for Applied Systems Analysis). 1995. *Global Energy Perspectives to 2050 and Beyond.* London, England: World Energy Council.

World Bank. 1994. *The Cost of Air Pollution Abatement*, Policy Research Working Paper No. 1398. Washington, D.C.: World Bank.

World Bank. 1997a. *Clear Water, Blue Skies: China's Environment in the New Century.* Washington, D.C.: World Bank.

World Bank. 1997b. *Surviving Success: Policy Reform and the Future of Industrial Pollution in China.* Washington, D.C.: World Bank.

World Bank. 1998. *China: A Strategy for International Assistance to Accelerate Renewable Energy Development.* Washington, D.C.: World Bank.

Yan Luguang. 1998. *Renewable Energy in China and Suggestion for Sino-American Cooperation.* Institute of Electrical Engineering, Chinese Academy of Sciences. (mimeo)

Yao Fusheng. 1998. *The Huge Energy Conservation Potential of Mechanical and Electrical Products.* Chinese Academy of Engineering. (mimeo)

Zhao Renkai. 1998. *Nuclear Power Development in China.* China National Nuclear Corporation. (mimeo)

Zheng Jianchao. 1998. *Power Technologies—the Bridge to Sustainable Development.* Chinese Academy of Engineering. (mimeo)

Zhou Fengqi. 1998a. *Long Term Energy Future of China.* Beijing: Energy Research Institute, State Development Planning Commission. (mimeo)

Zhou Fengqi. 1998b. *Energy Industry in China and Challenges for the 21st Century.* Beijing: Energy Research Institute, State Development Planning Commission. (mimeo)